PROSPECTS OF MATHEMATICAL SCIENCE

PROSPECTS OF MATHEMATICAL SCIENCE

Gakusyuin University, Tokyo
17-19 November 1986

Edited by

T Mitsui
Gakusyuin University
K Nagasaka
University of the Air
T Kano
Okayama University

World Scientific
Singapore • New Jersey • Hong Kong

Published by

World Scientific Publishing Co. Pte. Ltd.
P.O. Box 128, Farrer Road, Singapore 9128

U. S. A. office: World Scientific Publishing Co., Inc.
687 Hartwell Street, Teaneck NJ 07666, USA

PROSPECTS OF MATHEMATICAL SCIENCE

ISBN 9971-50-454-5

Printed in Singapore by JBW Printers & Binders Pte. Ltd.

FOREWORD

The scope of the symposium was to survey the recent development of each field of mathematical science, especially of one relating directly or indirectly to number theory. It seems that mathematical science has so far proceeded on with her trial different from that expected by D. Hilbert in 1900, to whom the prospect of her appeared so bright that all sciences were supposed to be unified into a form of mathematical science in the twentieth century. But who knows the truth behind the appearance even though this century is in the last stage?

If the symposium were significant somehow, it should be due to the assistance and efforts made from the beginning to the last by many people to whom we would like to express our sense of profound gratitude.

The Organizing Committee

EDITING PRINCIPLE

This Proceedings consist of 16 refereed papers, the authors of which are speakers and/or participants of the international symposium on Prospects of Mathematical Science. Each submitted paper is, in principle, reviewed by two referees and the editorial board determines the final decision of acceptance, revision or rejection based on the reports of two referees. The composition of the authors and their referees are expected to be international, that is, we set one referee outside of Japan for a Japanese author and vice versa.

We here would like to express our sincere thanks to all referees for their comments and corrections.

Dr. Kenji Nagasaka
Secretary of
The Proceedings of
The Prospects of
Mathematical Science

TABLE OF CONTENTS

Page

LIST OF PARTICIPANTS

OF

THE INTERNATIONAL SYMPOSIUM

ON

PROSPECTS OF MATHEMATICAL SCIENCE

17-19 NOVEMBER 1986 TOKYO

GAKUSHUIN UNIVERSITY

1. Norio Adachi, Fac. of Sci. and Engineering, Waseda Univ., Tokyo, Japan

2. Shiro Ando, College of Engineering, Hosei Univ., Tokyo, Japan

3. Kazuhiko Aomoto, Dept. of Math., Nagoya Univ., Nagoya, Japan

4. Akira Asada, Dept. of Math., Fac. of Sci., Shinshu Univ., Matsumoto, Japan

5. Shigeki Egami, Fac. of Liberal Arts, Toyama Univ., Toyama, Japan

6. Masahiko Fujiwara, Fac. of Sci., Ochanomizu Univ., Tokyo, Japan

7. Akio Fujii, Dept. of Math., Rikkyo Univ., Tokyo, Japan

8. Harry Furstenberg, Hebrew Univ., Jerusalem, Israel; Dept. of Math., Univ. Chicago, Illinois, U.S.A.

9. Takashi, Harase, Fac. of Sci., Tokyo Inst. of Technology, Tokyo, Japan

10. Tetsuya Hattori, Dept. of Physics, Fac. of Sci., Gakushuin Univ., Tokyo, Japan

11. Sin Hitotumatu, Research Inst. for Math. Sci., Kyoto Univ., Kyoto, Japan

12. Kuniaki Horie, Dept. of Math., Tokyo Metropolitan Univ., Tokyo, Japan

13. Mitsuko Horie, Dept. of Math., Tokyo Metropolitan Univ., Tokyo, Japan

14. Shunji Ito, Tsuda College, Tokyo, Japan

15. Yuji Ito, Fac. of Sci. and Technology, Keio Univ., Yokohama, Japan

16. Shigeru Kanemitsu, Dept. of Math., Fac. of Sci., Kyushu Univ., Fukuoka, Japan

17. Takeshi Kano, Dept. of Math., Fac. of Sci., Okayama Univ., Okayama, Japan

18. Yoshikazu Karamatsu, Utsunomiya Univ., Utsunomiya, Japan

19. Mayumi Kawauchi, Ochanomizu Univ., Tokyo, Japan

20. Norio Kono, Inst. of Math., Yoshida College, Kyoto Univ., Kyoto, Japan

21. Yukari Kosugi, Tsuda College, Tokyo, Japan

22. Toshiyuki Maebashi, Fac. of Sci., Kumamoto Univ., Kumamoto, Japan

23. Kohji Matsumoto, Dept. of Math., Fac. of Education, Iwate Univ., Morioka, Japan

24. Tsutomu Matsumoto, Division of Electrical and Computer Engineering, Fac. of Engineering, Yokohama National Univ., Yokohama, Japan

25. Jean-Loup Mauclaire, L'U.E.R. de Math. et Inf., Univ. Paris VII, Paris, France

26. Michel Mendes France, L'U.E.R. de Math. et Inf., Univ. Bordeaux I, Talence, France

27. M. Mishima, Gakushuin Univ., Tokyo, Japan

28. Takayoshi Mitsui, Dept. of Math., Fac. of Sci., Gakushuin Univ., Tokyo, Japan

29. Akio Miyai, Dept. of Math., Fac. of Education, Iwate Univ., Morioka, Japan

30. Katsuya Miyake, Dept. of Math., College of General Education, Nagoya Univ., Nagoya, Japan

31. Yasuo Morita, Math. Inst., Tohoku Univ., Sendai, Japan

32. Yoichi Motohashi, Dept. of Math., College of Sci. and Technology Nihon Univ., Tokyo, Japan

33. Leo Murara, Meiji Gakuin Univ., Tokyo, Japan

34. Kenji Nagasaka, Univ. of the Air, Chiba, Japan

35. Norikata Nakagoshi, Dept. of Math., Toyama Univ., Toyama, Japan

36. Toru Nakahara, Dept. of Math., Fac. of Sci. and Engineering, Saga Univ., Saga, Japan

37. Yoshinobu Nakai, Dept. of Math., Fac. of Education, Yamanashi Univ., Kofu, Japan

38. Shin Nakano, Gakushuin Univ., Tokyo, Japan

39. Masumi Nakajima, Dept. of Math., Rikkyo Univ., Tokyo, Japan

40. Hideaki Nakazawa, Fac. of Sci. and Technology, Keio Univ., Yokohama, Japan

41. Harald Niederreiter, Inst. of Math., Austrian Acad. of Sci., Vienna, Austria

42. Yoshitaka Odai, Tokyo Metropolitan Univ., Tokyo, Japan

43. Kosaku Okutsu, Gakushuin Univ., Tokyo, Japan

44. Hiroyuki Osada, Dept. of Math., Fac. of Sci., Rikkyo Univ., Tokyo, Japan

45. Yve-François Pêtermann, Section de Math., Univ. de Genêve, Genêve, Switzerland

46. Hiroshi Sugita, Dept. of Pure and Applied Sci., College of Arts and Sci., Univ. of Tokyo, Tokyo, Japan

47. Yoichiro Takahashi, Dept. of Pure and Applied Sci., College of Arts and Sci., Univ. of Tokyo, Tokyo, Japan

48. Akiyoshi Takamura, Rikkyo Univ., Tokyo, Japan

49. Jun-ichi Tamura, International Junior College, Tokyo, Japan

50. Shigeru Tanaka, Tsuda College, Tokyo, Japan

51. Tikao Tatuzawa, Ikutoku Technical Univ., Kanagawa, Japan

52. Shu Tezuka, Air Communication System Laboratory, Osaka, Japan

53. Isao Wakabayashi, Dept. of Math., Tokyo Univ. of Agriculture and Technology, Tokyo, Japan

54. Hirofumi Yamada, Hiroshima Univ., Hiroshima, Japan

55. Noriko Yamamoto, Ochanomizu Univ., Tokyo, Japan

56. Gosuke Yamano, Kanazawa Inst. of Technology, Kanazawa, Japan

57. Takeshi Yoshimoto, Dept. of Math., Toyo Univ., Kawagoe, Japan

58. Xu Guangshan, Inst. of Math., Academia Sinica, Beijing, People's Republic of China

PROGRAMME OF THE

INTERNATIONAL SYMPOSIUM

ON

PROSPECTS OF MATHEMATICAL SCIENCE

AT

CENTENNIAL MEMORIAL HALL OF

GAKUSHUIN UNIVERSITY

17 November 1986

10:00 - 10:15	Opening Ceremony

Chairman of Morning Session: T. Mitsui (Gakushuin Univ.)

10:30 - 11:30	M. Mendês France (Univ. Bordeaux I) Some Applications of the Theory of Automata

Chairman of Afternoon Session: T. Nakahara (Saga Univ.)

13:00 - 14:00	Xu Guangshan (Academia Sinica) On Siegel's G-functions
14:00 - 15:00	Y. Morita (Tohoku Univ.) A Lower Estimate of Lp(1,χ) for a Dirichlet Character
15:20 - 16:20	I. Wakabayashi (Tokyo Univ. of Agriculture and Technology) Aigebraic Values of Functions on the Unit Disk
16:20 - 17:20	Y.-F. Pêtermann (Univ. Genève) Divisor Problems and Exponent Pairs: On a Conjecture by Chowla and Walum

18 November 1986

Chairman of Morning Session: S. Ito (Tsuda College)

9:30 - 10:30	H. Furstenberg (Hebrew Univ.) Unconventional Ergodic Theorems in Number Theory
10:30 - 11:30	Y. Ito (Keio Univ.) On a Class of Ergodic Infinite Measure Preserving Transformations

Chairman of Afternoon Session: S. Hitotumatu (Kyoto Univ.)

13:00 - 14:00	J.-L. Mauclaire (Inst. Stat. Math.) Integration and Number Theory
14:00 - 15:00	T. Maebashi (Kumamoto Univ.) Dirac Monopoles and t'Hooft Monopoles
15:00 - 16:00	Y. Takahashi (Univ. of Tokyo) On Asymptotic Behavior of Measures of Small Tubes for Large Time
18:00 - 20:00	Welcome Party at Taisho Central Hotel

19 November 1986

Chairman of Morning Session: S. Ando (Hosei Univ.)

9:30 - 10:30	H. Niederreiter (Austrian Acad. Sci.) Cryptology - The Mathematical Theory of Data Security
10:30 - 11:30	T. Tatuzawa (Ikutoku Technical Univ.) On a Lower Bound for $L(1,\chi)$

Chairman of Afternoon Session: K. Nagasaka (Shinshu Univ.)

12:30 - 13:30	Y. Motohashi (Nihon Univ.) On the Mean Square of Dirichlet L-functions
13:30 - 14:30	S. Ito (Tsuda College) Ergodic Approach on Diophantine Approximation
14:30 - 15:30	K. Matsumoto (Rikkyo Univ.) A Mean-value Formula of the Riemann Zeta-function in the Critical Strip

PROSPECTS OF MATHEMATICAL SCIENCE

Proc. Prospects
of Math. Sci.
World Sci. Pub.
1-12, 1988

ALGEBRAIC EQUATIONS FOR GREEN KERNEL ON A FREE GROUP

KAZUHIKO AOMOTO

Department of Mathematics
Nagoya University
Chikusa-ku, Nagoya 464
Japan

A way of explicit construciton of Green kernel is presented for a linear difference operator of nearest neighbours on a free group which is left-invariant with respect to a subgroup of finite index. This is an extension of a result about spectra in [1] or [9] for convolution operators on a free group.

The algebraicity of Green kernel is a most crucial fact for this. W. Woess has proved in [10] its algebraicity under most general context. See [4],[9] and [11] for related topics.

1. STRUCTURE EQUATIONS FOR GREEN KERNELS

Let Γ be a free group of g generators $\{\sigma_1,\cdots,\sigma_g\}$ and Γ_0 be a subgroup of finite index N. We consider a linear operator A on $\ell^2[\Gamma]$ which is left-invariant with respect to Γ_0:

$$Au(\gamma) = \sum_{\gamma'\in\Gamma} a_{\gamma,\gamma'}\, u(\gamma'), \quad \gamma\in\Gamma , \tag{1.1}$$

for $u(\gamma)\in\ell\,[\Gamma]$ and $a_{\gamma,\gamma'}\in\mathbb{R}$. We assume that:

1) A is symmetric, namely $a_{\gamma,\gamma'} = a_{\gamma',\gamma}$ for $\gamma,\gamma'\in\Gamma$.

2) A is of nearest neighbours, namely $a_{\gamma,\gamma'} = 0$,

provided that $\mathrm{dis}(\gamma,\gamma') \geq 2$, where $\mathrm{dis}(\gamma,\gamma')$ denotes the length of the minimal reduced expression of $\gamma^{-1}\gamma'$.

3) $a_{\gamma,\gamma'} \neq 0$ for $\mathrm{dis}(\gamma,\gamma') = 1$.

4) $a_{\gamma_0\gamma,\gamma_0\gamma'} = a_{\gamma,\gamma'}$ for arbitrary $\gamma_0 \in \Gamma_0$.

Then it is known that A is a self-adjoint bounded operator so that for $z \in \mathbf{C}$ such that $\mathrm{Im}\, z \neq 0$, there exists the unique Green kernel $G(\gamma,\gamma'|z)$ which defines the resolvent $(z - A)^{-1}$ on $\ell^2[\Gamma]$.

We have proved in the previous paper [1] (see also [9]) that the Green kernel $G(\gamma,\gamma'|z)$ always becomes an algebraic function of z. The purpose of this note is to give explicit equations *in terms of the diagonal elements* $G(\gamma,\gamma|z)$ $\gamma \in \Gamma$ *of it*. We denote by W_γ the inverse of $G(\gamma,\gamma|z)$. Then we have $W_{\gamma_0\gamma} = W_\gamma$ for arbitrary $\gamma_0 \in \Gamma_0$, so that the quantity W_γ is defined on the quotient $\Gamma_0 \backslash \Gamma$.

Let $d\theta(\gamma,\gamma'|\lambda)$ be the spectral kernel for A such that

$$a_{\gamma,\gamma'} = \int_{-\infty}^{\infty} \lambda \; d\theta(\gamma,\gamma'|\lambda) , \tag{1.2}$$

$$\delta_{\gamma,\gamma'} = \int_{-\infty}^{\infty} \lambda \; d\theta(\gamma,\gamma'|\lambda) , \tag{1.3}$$

where $\delta_{\gamma,\gamma'}$ denotes the Kronecker's delta. Then from the integral representation:

$$G(\gamma,\gamma|z) = \int_{-\infty}^{\infty} \frac{d\theta(\gamma,\gamma|\lambda)}{z-\lambda} , \tag{1.4}$$

we see that $W_\gamma(z)$ is holomorphic and never vanishes for $\mathrm{Im}\, z \neq 0$. In fact we have, from a general property of the Green kernel,

$$\mathrm{Im}\, W_\gamma(z) \cdot \mathrm{Im}\, z > 0 \quad \text{for } \mathrm{Im}\, z \neq 0. \tag{1.5}$$

This also follows from Corollary 1 of Proposition 1 through a direct computation.

First we want to prove the following theorem:

Theorem 1. i) *The* N *quantities* $W_\gamma(z)$, $\gamma \in \Gamma_0 \backslash \Gamma$, *satisfy the following* N*-algebraic equations*:

$$z - a_{\gamma,\gamma} - W_\gamma$$
$$= \sum_{\langle\gamma,\gamma'\rangle} \frac{1}{2} \left(- W_\gamma + \sqrt{W_\gamma^2 + 4a^2_{\gamma,\gamma'} W_\gamma / W_{\gamma'}}\right) , \tag{E}$$

where $\langle\gamma,\gamma'\rangle$ *means that* $\gamma' \in \Gamma$ *runs over all elements adjacent to* γ.

ii) *We denote by* $\alpha_i^{(+)}(\gamma|z)$ *and* $\alpha_i^{(-)}(\gamma|z)$ *the quotients* $G(\gamma\sigma_i, e|z)/G(\gamma,e|z)$ *for* $\gamma < \gamma\sigma_i$ *i.e.* $\mathrm{dis}(e,\gamma\sigma_i) = \mathrm{dis}(e,\gamma)+1$ *and* $G(\gamma\sigma_i^{-1}|z)/G(\gamma,e|z)$ *for* $\gamma < \gamma\sigma_i^{-1}$ *i.e.* $\mathrm{dis}(e,\gamma\sigma_i^{-1}) = \mathrm{dis}(e,\gamma)+1$, *respectively. Then we have*

$$\alpha_i^{(+)}(\gamma|z)(\text{or } \alpha_i^{(-)}(\gamma|z)) = \frac{-W_\gamma + \sqrt{W_\gamma^2 + 4a^2_{\gamma,\gamma'} W_\gamma / W_{\gamma'}}}{2a_{\gamma,\gamma'}} , \tag{1.6}$$

for $\gamma' = \gamma\sigma_i$ (*and* $\gamma\sigma_i^{-1}$ *respectively*), *such that*

$$\alpha_i^{(\pm)}(\gamma|z) = \frac{a_{\gamma,\gamma'}}{z} + O(1) \quad \text{for } z \to \infty. \tag{1.7}$$

Proof of Theorem 1. The Green kernel $G(\gamma,\gamma_0|z)$ satisfies the following equations:

$$(z-a_{\gamma,\gamma}) \cdot G(\gamma,\gamma_0|z) - \sum_{\langle\gamma,\gamma'\rangle} a_{\gamma,\gamma'} G(\gamma',\gamma_0|z) = \delta_{\gamma,\gamma_0}. \tag{1.8}$$

As we have shown in [1], the quotients $G(\gamma\sigma_i^{\pm 1},\gamma_0|z)/G(\gamma,\gamma_0|z)$ are independent of γ_0 provided $\gamma_0^{-1}\gamma < \gamma_0^{-1}\gamma\sigma_i$ or $\gamma_0^{-1}\gamma\sigma_i^{-1}$, and are equal to $\alpha_i^{(\pm)}(\gamma|z)$ respectively. Suppose that $\gamma_0^{-1}\gamma$ has an expression $(\gamma_0^{-1}\gamma')\cdot\sigma_k^{\pm 1}$ such that $\gamma_0^{-1}\gamma' < \gamma_0^{-1}\gamma$ for some k, $1 \leq k \leq g$, then (1.8) is reexpressed in terms of $\alpha_i^{(\pm)}(\gamma|z)$ as follows:

$$z - a_{\gamma,\gamma} - \sum_{j\neq k} \{a_{\gamma,\gamma\sigma_j}\alpha_j^{(+)}(\gamma|z) + a_{\gamma,\gamma\sigma_j^{-1}}\alpha_j^{(-)}(\gamma|z)\}$$
$$- a_{\gamma,\gamma\sigma_k^{\pm 1}}\, \alpha_k^{(\pm)}(\gamma|z) - \frac{a_{\gamma,\gamma\sigma_k^{\mp 1}}}{\alpha_k^{(\pm)}(\gamma\sigma_k^{\mp 1}|z)} = 0 . \tag{1.9}$$

On the other hand, by putting $\gamma = \gamma_0$ in (1.8) we have

$$W_\gamma = z - a_{\gamma,\gamma} - \sum_{j=1}^{g} \{a_{\gamma,\gamma\sigma_j} \alpha_j^{(+)}(\gamma|z) + a_{\gamma,\gamma\sigma_j^{-1}} \alpha_j^{(-)}(\gamma|z)\} . \quad (1.10)$$

These two imply

$$W_\gamma = a_{\gamma,\gamma\sigma_k^{\mp 1}} \left(\frac{1}{\alpha_k^{(\pm)}(\gamma\sigma_k^{\mp 1}|z)} - \alpha_k^{(\mp)}(\gamma|z) \right) . \quad (1.11)$$

In the same manner we choose γ_0 such that $\gamma_0^{-1}\gamma > \gamma_0^{-1}\gamma\sigma_k^{-1}$ or $\gamma_0^{-1}\gamma > \gamma_0^{-1}\gamma\sigma_k$ and, by reversing the role of γ and $\gamma'=\gamma\sigma_k^{\mp 1}$, we have

$$W_{\gamma\sigma_k^{\mp 1}} = a_{\gamma\sigma_k^{\mp 1},\gamma} \left(\frac{1}{\alpha_k^{(\mp)}(\gamma|z)} - \alpha_k^{(\pm)}(\gamma\sigma_k^{\mp 1}|z) \right) . \quad (1.12)$$

Namely if $\gamma_0^{-1}\gamma = (\gamma_0^{-1}\gamma')\sigma_k$ (reduced expression), then

$$\begin{aligned} W_\gamma &= a_{\gamma,\gamma'} \left(\frac{1}{\alpha_k^{(+)}(\gamma'|z)} - \alpha_k^{(-)}(\gamma|z) \right) , \\ W_{\gamma'} &= a_{\gamma',\gamma} \left(\frac{1}{\alpha_k^{(-)}(\gamma|z)} - \alpha_k^{(+)}(\gamma'|z) \right) , \end{aligned} \quad (1.13)$$

and if $\gamma_0^{-1}\gamma = (\gamma_0^{-1}\gamma')\sigma_k^{-1}$ (reduced expression), then

$$\begin{aligned} W_\gamma &= a_{\gamma,\gamma'} \left(\frac{1}{\alpha_k^{(-)}(\gamma'|z)} - \alpha_k^{(+)}(\gamma|z) \right) , \\ W_{\gamma'} &= a_{\gamma',\gamma} \left(\frac{1}{\alpha_k^{(+)}(\gamma|z)} - \alpha_k^{(-)}(\gamma'|z) \right) , \end{aligned} \quad (1.14)$$

respectively. By these equalities, we can solve for $\alpha_k^{(\pm)}(\gamma|z)$, $\alpha_k^{(\pm)}(\gamma'|z)$ as in (1.6). The substitution of these formulae into the equations (1.10) gives the desired one ($\mathfrak{E}$).

(1.6) can also be expressed in the following manner:

$$2a_{\gamma,\gamma'} G(\gamma,\gamma'|z) = -1 + \sqrt{1 + 4a_{\gamma,\gamma'}^2 G(\gamma,\gamma|z)G(\gamma'\gamma'|z)} , \quad (1.15)$$

for $\gamma' = \gamma\sigma_k^{\pm 1}$.

We denote by $\mathfrak{H}_{\pm}$ the upper or lower half plane $\mathrm{Im}\, z > 0$ or $\mathrm{Im}\, z < 0$ and by $\overline{\mathfrak{H}}_{\pm}$ the closures of them in $\mathbf{CP}^1$ respectively. Then (1.5) and the above formulae show.

Corollary of Theorem 1. *No* $\alpha_k^{(\pm)}(\gamma|z)$ *either vanish or become infinite in each* $\mathfrak{H}_+$ *and* $\mathfrak{H}_-$. *Hence* $\sqrt{W_\gamma^2 + 4a_{\gamma,\gamma'}^2 \frac{W_\gamma}{W_{\gamma'}}}$ *which are equal to* $a_{\gamma,\gamma'} \left(\frac{1}{\alpha_k^{(\mp)}(\gamma'|z)} + \alpha_k^{(\pm)}(\gamma|z)\right)$ *are holomorphic there.*

2. SOME PROPERTIES OF MULTIPLIERS

For $z \in \mathfrak{H}_+ \cup \mathfrak{H}_-$, we put $S_j^{(\pm)}(\gamma|z) = a_{\gamma,\sigma_j^{\mp 1}}$, $(\alpha_j^{(\pm)}(\gamma|z) - \alpha_j^{(\pm)}(\gamma|\bar{z}))$ and $s_j^{(\pm)}(\gamma|z) = \frac{1}{i} S_j^{(\pm)}(\gamma|z)$ respectively. Then

Proposition 1.

$$\mathrm{Im}\, S_j^{(\pm)}(\gamma|z) \cdot \mathrm{Im}\, z < 0. \tag{2.1}$$

To prove Proposition 1 we need the following lemma.

Lemma 1. *Let* δ *be* $\mathrm{Im}\, z$. *Then for* $\gamma' = \gamma\sigma_k^{\mp 1}$ *reduced expression,*

$$\frac{1}{i}(W_\gamma(z) - W_\gamma(\bar{z})) = 2\delta - \sum_{j=1}^{g} (s_j^{(+)}(\gamma|z) + s_j^{(-)}(\gamma|z))$$

$$= - s_k^{(\mp)}(\gamma|z) - \frac{s_k^{(\pm)}(\gamma'|z)}{|\alpha_k^{(\pm)}(\gamma'|z)|^2}\,. \tag{2.2}$$

Proof. The first identity follows from (1.10). The second one follows from the substraction between each of (1.11) $\sim$ (1.14) and its complex conjugate. From (2.2) we have for $\gamma' = \gamma\sigma^{(-\varepsilon)}$ for $\varepsilon = \pm 1$

$$s_k^{(\varepsilon)}(\gamma|z) = 2\delta - \sum_{j\neq k}(s_j^{(+)}(\gamma|z) + s_j^{(-)}(\gamma|z)) + \frac{s_k^{(\varepsilon)}(\gamma'|z)}{|\alpha_k^{(\varepsilon)}(\gamma'|z)|^2} . \qquad (2.3)$$

Let n be the smallest positive integer such that $(\sigma_k^{(-\varepsilon)})^n \in \Gamma_0$. We denote by $\gamma^{(\nu)} = \gamma\cdot(\sigma_k^{(-\varepsilon)})^\nu$ for $0 \leqq \nu \leqq n$. Since $\gamma^{(n)} \equiv \gamma \mod \Gamma_0$, $\alpha_k^{(\varepsilon)}(\gamma|z) = \alpha_k^{(\varepsilon)}(\gamma^{(n)}|z)$, whence $s_k^{(\varepsilon)}(\gamma|z) = s_k^{(\varepsilon)}(\gamma^{(n)}|z)$. Applying (2.3) to each $\gamma^{(\nu)}$ we obtain

$$2\delta - \sum_{j\neq k}(s_j^{(-)}(\gamma^{(\nu)}|z) + s_j^{(+)}(\gamma^{(\nu)}|z)) + \frac{s_k^{(\varepsilon)}(\gamma^{(\nu+1)}|z)}{|\alpha_k^{(\varepsilon)}(\gamma^{(\nu+1)}|z)|^2} = s_k^{(\varepsilon)}(\gamma^{(\nu)}|z) . \qquad (2.4)$$

Hence

$$s_k^{(\varepsilon)}(\gamma|z) = \sum_{\nu=0}^{n-1} \frac{2\delta - \sum_{j\neq k}(s_j^{(+)}(\gamma^{(\nu)}|z) + s_j^{(-)}(\gamma^{(\nu)}|z))}{|\alpha_k^{(\varepsilon)}(\gamma^{(1)}|z)|^2 \cdots |\alpha_k^{(\varepsilon)}(\gamma^{(\nu)}|z)|^2} + \frac{s_k^{(\varepsilon)}(\gamma^{(n)}|z)}{|\alpha_k^{(\varepsilon)}(\gamma^{(1)}|z)|^2 \cdots |\alpha_k^{(\varepsilon)}(\gamma^{(n)}|z)|^2} , \qquad (2.5)$$

namely

$$\{1 - \frac{1}{|\alpha_k^{(\varepsilon)}(\gamma^{(1)}|z)|^2 \cdots |\alpha_k^{(\varepsilon)}(\gamma^{(n)}|z)|^2}\} s_k^{(\varepsilon)}(\gamma|z) = \sum_{\nu=0}^{n-1} \frac{2\delta - \sum_{j\neq k}(s_j^{(+)}(\gamma^{(\nu)}|z) + s_j^{(-)}(\gamma^{(\nu)}|z))}{|\alpha_k^{(\varepsilon)}(\gamma^{(1)}|z)|^2 \cdots |\alpha_k^{(\varepsilon)}(\gamma^{(\nu)}|z)|^2} . \qquad (2.6)$$

Proof of Proposition 1. We may assume that $z \in \mathbb{H}_+$. We denote by Ω the set of all points $z \in \mathbb{H}_+$ such that $s_j^{(\pm)}(\gamma|z) < 0$, for all $\gamma \in \Gamma$ and j. Then Ω is an open set in $\mathbb{H}_+$. We want to show that Ω coincides with $\mathbb{H}_+$ itself. Assume that $\Omega \subsetneqq \mathbb{H}_+$. Let z_0 be a point $\bar{\Omega} - \Omega$ in $\mathbb{H}_+$. Then obviously $s_j^{(\pm)}(\gamma|z) \leqq 0$ for all $\gamma \in \Gamma$ and j, and $s_k^{(\varepsilon)}(\gamma|z) = 0$ for at least one of γ, k and ε. We fix them and apply the formula (2.6) to $s_k^{(\varepsilon)}(\gamma|z)$. The right hand side is larger than or equal to $2\delta > 0$, while the left hand side is equal to 0. This is a contradiction. This implies $\bar{\Omega} = \Omega$ and therefore Ω must coincide with $\mathbb{H}_+$.

Remark. From (2.6) we have for $z \in \mathbb{H}_+$

$$|\alpha_k^{(\varepsilon)}(\gamma^{(1)}|z)| \cdots |\alpha_k^{(\varepsilon)}(\gamma^{(n)}|z)| < 1 . \tag{2.7}$$

Indeed otherwise, since $s_j^{(\varepsilon)}(\gamma|z) < 0$ by Proposition 1, the left hand side of (2.6) is non-positive, while the right hand side is at least equal to 2δ, which is a contradiction.

Corollary of Proposition 1.

$$\frac{z - \bar{z}}{W_\gamma(z) - W_\gamma(\bar{z})} < 1 . \tag{2.8}$$

Let $\mathfrak{C} \subset \mathbf{C} \times (\mathbf{C}^*)^N$ be the algebraic curve defined by $(\mathbf{E})$. Then Theorem 1 shows that the mappings Φ and Φ_0:

$$\begin{array}{ccc} z \in \mathbb{H}_+ \cup \mathbb{H}_- & \xrightarrow{\ \Phi\ } & (z, W_{\gamma_1}(z), \cdots, W_{\gamma_N}(z)) \in \mathfrak{C} \\ & \searrow^{\Phi_0} & \downarrow \rho \\ & & (W_{\gamma_1}(z), \cdots, W_{\gamma_N}(z)) \in (\mathbf{C}^*)^N \end{array} \tag{2.9}$$

where ρ denotes the canonical projection, are well defined. We denote by $\mathfrak{C}_\pm = \Phi(\mathbb{H}_\pm)$ respectively. In view of (1.5) $\Phi_0(\mathbb{H}_+)$ and $\Phi_0(\mathbb{H}_-)$ are contained in $(\mathbb{H}_+)^N$ and $(\mathbb{H}_-)^N$ respectively. It seems probable that the following holds.

Conjecture 1. *Φ_0 is univalent from each $\mathbb{H}_+$ or $\mathbb{H}_-$.*

This conjecture turns out to be true in the case where g=1 or N=1 (see Examples 1 and 2).

Conjecture 2. *The boundary points of $\sigma(A)$ coincides with the critical points z of the mapping*

$$\Phi_0^{-1} : (W_{\gamma_1}(z),\cdots,W_{\gamma_N}(z)) \in \overline{\mathfrak{G}}_+ \to z \in \mathbb{H}_+ , \qquad (2.10)$$

where $\overline{\mathfrak{G}}_+$ *denotes the closure of* $\mathfrak{G}_+$ *in* $\mathfrak{G}$.

We denote by γ_j, $1 \leq j \leq N$, a complete set of representatives of $\Gamma_0 \backslash \Gamma$. Since the equations $(\mathfrak{E})$ give the entries $a_{\gamma,\gamma}$ and $a_{\gamma,\gamma'}$ for $\langle\gamma,\gamma'\rangle$ in a unique way, Theorem 1 implies that the N functions $G(\gamma_j,\gamma_j|z)$ completely determine the operator A except for signs of elements. The signs are determined by (1.6).

One may ask finally the following question: *Is* A *determined by the mean value* $\frac{1}{N}\sum_{j=1}^{N} G(\gamma_j,\gamma_j|z)$ *up to unitary equivalence?*

3. THE CONVERSE OF THEOREM 1

Theorem 2. *We can define* $\tilde{W}_\gamma(z)$, $\gamma \in \Gamma_0 \backslash \Gamma$ *in a unique way such that* $W_\gamma = \tilde{W}_\gamma(z)$ *are algebraic functions of* z *satisfying* $(\mathfrak{E})$ *and having Laurent expansions at* $z = \infty$:

$$\tilde{W}_\gamma(z) = z + O\left(\frac{1}{|z|} \right) . \qquad (3.1)$$

$\tilde{W}_\gamma(z)$ *can be holomorphically extended into the domains* $\mathbb{H}_+ \cup \mathbb{H}_-$ *and satisfy the inequality*

$$\tilde{W}_\gamma(z) \cdot \mathrm{Im}\, z > 0 . \qquad (3.2)$$

$\tilde{W}_\gamma(z)$ *conicides with* $W_\gamma(z)$ *in* $\mathbb{H}_+ \cup \mathbb{H}_-$.

Proof. We first assume $|\mathrm{Im}\, z| > M$ for a sufficiently large number M. Then we can solve $(\mathfrak{E})$ in a unique way such that the solutions $W_\gamma = \tilde{W}_\gamma$ satisfy (3.1). We define $\alpha_k^{(\pm)}(\gamma|z)$ and $s_k^{(\pm)}(\gamma|z)$ by the formulae (1.6) and as in 2., replacing $W_\gamma(z)$ by $\tilde{W}_\gamma(z)$, so that (1.11)$\sim$(1.14) hold. Hence $(\mathfrak{E})$ implies (1.9). $G(\gamma,\gamma'|z)$ are defined inductively on the integer $\mathrm{dis}(\gamma,\gamma'|z)$ beginning from $G(\gamma,\gamma|z) = \frac{1}{W_\gamma(z)}$ through the formulae $G(\gamma,\gamma'\sigma_i^{\pm 1}|z) = G(\gamma,\gamma'|z)\,\alpha_i^{(\pm)}(\gamma'|z)$ for $\gamma^{-1}\gamma'\sigma_i^{\pm 1} > \gamma^{-1}\gamma'$. As a result we have the equations (1.8) which means that the functions thus defined $G(\gamma,\gamma'|z)$ are the same as the elements of the Green kernel for A, because the latter is uniquely determined by $(\mathcal{E})$ and the asymptotic behaviours at the infinity of z. Once the functions $G(\gamma,\gamma'|z)$ coincide with the Green kernel for A, they are holomorphic in $\mathbb{H}_+ \cup \mathbb{H}_-$. They are also algebraic in z.

Remark. Perhaps one may prove from $(\mathcal{E})$ by a direct argument that the functions $\tilde{W}_\gamma(z)$ can be continued analytically in $\mathbb{H}_+ \cup \mathbb{H}_-$. But we haven't been able to do it.

4. EXAMPLES

1) Periodic Jacobi matrices. In this case $\Gamma = \mathbf{Z}$ and $\Gamma_0 = N\mathbf{Z}$ for an integer $N>0$. A becomes a periodic Jacobi-matrix $((a_{n,m}))_{n,m\in\mathbf{Z}}$ such that $a_{n+N,m+N} = a_{n,m}$. $(\mathcal{E})$ becomes as follows:

$$z - a_{n,n} = \frac{1}{2}\left\{\sqrt{W_n^2 + 4a_{n,n+1}^2 \frac{W_n}{W_{n+1}}} + \sqrt{W_n^2 + 4a_{n,n-1}^2 \frac{W_n}{W_{n-1}}}\right\} . \tag{4.1}$$

By using Floquet multiplier $h = G(n,n+N|z)/G(n,n|z)$, the n-shifted characteristic equation can be expressed as follows:

$$h + h^{-1} - \frac{P(z)}{a_{0,1}a_{1,2}\cdots a_{N-1,N}} = 0 , \tag{4.2}$$

where $P(z)$ is a polynomial of degree N and independent of n. Then it is well-known (see [8] p.123-126) that

$$G(n,n|z) = \frac{\Delta_{n,n}(z)}{\sqrt{P(z)^2 - 4a_{0,1}^2\cdots a_{N-1,N}^2}} , \tag{4.3}$$

where $\Delta_{n,n}(z)$ denotes the (n,n)-th cofactor of the above matrix and becomes a polynomial of degree N-1. Since $\Delta_{j,j}(z)$ are coprime to each other, there exist $\lambda_0,\cdots,\lambda_{N-1}, \mu_0,\cdots,\mu_{N-1}\in\mathbb{R}$ such that

$$\sum_{j=0}^{N-1} \lambda_j \, \Delta_{j,j}(z) = 1 \, ,$$

$$\sum_{j=0}^{N-1} \mu_j \, \Delta_{j,j}(z) = z \, . \tag{4.4}$$

Hence we have

$$z = \frac{\sum_{j=0}^{N-1} \mu_j \, G(j,j|z)}{\sum_{j=0}^{N-1} \lambda_j \, G(j,j|z)} \, . \tag{4.5}$$

In other words z is uniquely determined by the values of $G(j,j|z)$, so that Conjectures 1 and 2 are true. We put

$$\mathcal{G}_n = \sqrt{1 + \frac{4a_{n,n+1}^2}{W_n \, W_{n+1}}} \quad \text{and} \quad \psi_n = \frac{1}{W_n} \, . \tag{4.6}$$

Then $(\mathcal{E})$ is equivalent to the following system of equations:

$$a_{n,n-1}^2 \, \psi_{n-1} = (z-a_{n,n})^2 \, \psi_n - (z-a_{n,n})\mathcal{G}_n + a_{n,n+1} \, \psi_{n+1} \, ,$$

$$2(z-a_{n,n})\psi_n = \mathcal{G}_n + \mathcal{G}_{n-1} \, . \tag{4.7}$$

By eliminating $\mathcal{G}_n$ from these two equations, we have the linear difference equation of 3rd order in $\psi_n = G(n,n|z)$:

$$\frac{-a_{n-1,n-2}^2 \cdot \psi_{n-2} + (z-a_{n-1,n-1})^2 \cdot \psi_{n-1} + a_{n-1,n}^2 \cdot \psi_n}{z - a_{n-1,n-1}}$$

$$= \frac{a_{n,n-1}^2 \cdot \psi_{n-1} + (z-a_{n,n})^2 \cdot \psi_n + a_{n,n+1}^2 \cdot \psi_{n+1}}{z - a_{n,n}} \, . \tag{4.8}$$

R. Johnson and J. Moser (see [6] p.428) have shown that the continuous analogues of (4.7) and (4.8) play an important role in deducing K-d-V hierachies in the case of almost periodic potentials. It is an easy thing to establish similar results in our periodic case by using (4.7), (4.8) and a result in [8] p.123-126. Furthermore it seems interesting to ask if this result can be extended to the case of free groups.

2) Left invariant case, i.e. $N = 1$. In this case $\Gamma_0 = \Gamma$ and W_γ are all equal. Hence $(\mathcal{E})$ is reduced to the following:

$$z - W_e - a_{e,e} = \sum_{i=1}^{g} (- W_e + \sqrt{W_e^2 + 4a_{e,\sigma_i}^2}) . \qquad (4.9)$$

By using this formula, the spectrum and the eigen-function expansion have been investigated in [1] and [9]. It has been proved there that Conjectures 1 and 2 are true.

3) We assume that Γ_0 is the subgroup consisting of elements γ having reduced expressions: $\gamma = \sigma_{i_1}^{\varepsilon_1} \cdots \sigma_{i_n}^{\varepsilon_n}$ such that the sum of exponents of σ_1 contained in γ vanishes mod N. We take as the representatives of the coset space $\Gamma_0 \Gamma$, the set $\{e,\sigma_1,\sigma_1^2,\cdots,\sigma_1^{N-1}\}$ and put $W_n = W_{\sigma_1^n}$. Then $(\mathcal{E})$ is written as follows:

$$\begin{aligned} z - a_{n,n} - W_n = {} & \tfrac{1}{2}(-W_n + \sqrt{W_n^2 + 4a_{n,n+1}^2 \frac{W_n}{W_{n+1}}}) \\ & + \tfrac{1}{2}(-W_n + \sqrt{W_n^2 + 4a_{n,n-1}^2 \frac{W_n}{W_{n-1}}}) \\ & + \sum_{j\neq 1}^{g} \tfrac{1}{2}(-W_n + \sqrt{W_n^2 + 4(b_{n,j}^{(+)})^2}) \\ & + \sum_{j\neq 1}^{g} \tfrac{1}{2}(-W_n + \sqrt{W_n^2 + 4(b_{n,j}^{(-)})^2}) , \end{aligned} \qquad (4.10)$$

for $a_{n,n} = a_{\sigma_1^n,\sigma_1^n}$, and $a_{n,n\pm1} = a_{\sigma_1^n,\sigma_1^{n\pm1}}$ and $b_{n,j}^{(\pm)} = a_{\sigma_1^n,\sigma_1^n\sigma_j^{\pm1}}$.

Remark. We may consider a case where Γ is not necessarily a free group but a tree. If Γ is a finite tree, then the inverse $W_\gamma(z)$ of the function $G(\gamma,\gamma|z)$ satisfies $(\mathcal{E})$. Then it is a trivial fact that $W_\gamma(z)$ all turn out to be rational in z. Moreover *it seems strongly probable that the equations* (E) *have the unique solution* $W_\gamma(z)$ *which gives the inverse of* $G(\gamma,\gamma|z)$ *in case of considerably wide classes*, provided that we set a suitable boundary condition for $W_\gamma(z)$ at the infinite or finite boundary pionts of Γ.

REFERENCES

[1] Aomoto, K.: Spectral theory on a free group and algebraic curves, J. Fac. Sci. Univ. of Tokyo, 31, 297-318 (1984).

[2] Aomoto, K.: A formula of eigen-function expansion I, Case of asymptotic trees, Proc. Japan Acad., 61, 11-14 (1985).

[3] Aomoto, K. and Kato, Y.: Green functions and spectra on a free group of cyclic groups, submitted to Ann. Inst. Fourier.

[4] Cartwright, D.I. and Soardi, P.M.: Random walks on a free products, quotients and amalgams, Nagoya Math. J., 102, 163-180 (1986).

[5] Figã-Talamanca and Picardello, M., Harmonic Analysis on Free Groups, Lecture Notes in Pure Applied Math., 87, Marcel Dekker, 1983.

[6] Johnson, R. and Moser, J.: The rotation number for almost periodic potentials, Comm. Math. Phys., 84 (1982).

[7] Kato, Y.: Mixed periodic Jacobi continued fractions, Nagoya Math. J., 104, 129-148 (1986).

[8] Van Moerbeke, P. and Munford, D.: The spectrum of difference operators and algebraic curves, Acta, Math., 143, 93-154 (1979).

[9] Steger, T., Harmonic Analysis for an Anisotropic Random Walk on a Homogeneous Tree, Thesis, Washington Univ., St. Louis, 1985.

[10] Woess, W.: Context-free languages and random walks on groups, preprint, 1986, Inst. fur Math. und Angew. Geometrie, Austria.

[11] Woess, W.: Nearest neighbour random walks on free products of discrete groups, Bollettino U.M.I., 5-B, 961-982 (1986).

Proc. Prospects
of Math. Sci.
World Sci. Pub.
13-40, 1988

NON ABELIAN DE RHAM THEORY

AKIRA ASADA

Department of Mathematics
Faculty of Science
Shinshu University
Matsumoto, Nagano Pref.
Japan

1. INTRODUCTION

The main object of this paper is to state definitions and properties of the third non abelian de Rham sets. They are chiral and seem to relate to background geometry of string field theory. First we give rough definition and historical remarks of non abelian de Rham theory in Introduction.

Let M be a smooth manifold and G be a Lie group with the Lie algebra $\mathfrak{G}$. For simplicity, we assume throughout this paper $G=GL(n,C)$. We use the following notations:

G_t, G_d; the sheaves of germs of constant and smooth G-valued functions over M, respectively.

$\mathfrak{m}^1_R$, $\mathfrak{m}^1_L$; the sheaves of germs of matrix valued 1-forms θ such that $d\theta+\theta\wedge\theta=0$ and $d\theta-\theta\wedge\theta=0$, respectively.

G_* and $\mathfrak{m}^1$ mean either of G_t or G_d and $\mathfrak{m}^1_R$ or $\mathfrak{m}^1_L$.

By definitions and Frobenius' theorem, we have the following exact sequences of sheaves.

$$0\to G_t\to G_d\xrightarrow{\rho_R}\mathfrak{m}^1_R\to 0,\quad \rho_R(g)=g^{-1}dg,$$

$$0\to G_t\to G_d\xrightarrow{\rho_L}\mathfrak{m}^1_L\to 0,\quad \rho_L(g)=(dg)g^{-1}.$$

If the (p-1)-th cohomology set $H^{p-1}(M,\mathfrak{m}^1)$ of M with coefficients in $\mathfrak{m}^1$ is defined, then we call it *the p-th non abelian de Rham set of* M (with respect to G). *The p-th non abelian de Rham theory* studies the p-th non abelian de Rham set. The p-th non abelian de Rham set has an expression by p-forms. The correspondence from a representing cocycle of an element of $H^{p-1}(M,\mathfrak{m}^1)$ to its p-form expression is called *the de Rham correspondence*.

$H^0(M,\mathfrak{m}^1)$ was called the non abelian de Rham set of M by Oniscik and Gaveau ([12],[16],[17],[18]). Gaveau and Andersson gave its Hodge theory ([2],[12]). $H^1(M,\mathfrak{m}^1)$ was defined in 1978 ([3], cf.[1]). Its de Rham correspondence was given in 1983 ([4], cf.[21]). In these studies, only $\mathfrak{m}^1_R$ (or $\mathfrak{m}^1_L$) were used.

The definition of $H^2(M,\mathfrak{m}^1)$ and the following *chiral exact sequence* of non abelian cohomology sets are obtained in 1986. These lead us to use both of $\mathfrak{m}^1_R$ and $\mathfrak{m}^1_L$.

$$0 \to H^0(M,G_t) \to H^0(M,G_d) \xrightarrow{\rho_R} H^0(M,\mathfrak{m}^1_R) \xrightarrow{\delta_R} H^1(M,G_t) \to$$
$$0 \to H^0(M,G_t) \to H^0(M,G_d) \xrightarrow{\rho_L} H^0(M,\mathfrak{m}^1_L) \xrightarrow{\delta_L} H^1(M,G_t) \to$$
$$\to H^1(M,G_d) \xrightarrow{\rho_R{}^*} H^1(M,\mathfrak{m}^1_R) \xrightarrow{\delta_R} H^2(M,G_t)_R \to$$
$$\to H^1(M,G_d) \xrightarrow{\rho_L{}^*} H^1(M,\mathfrak{m}^1_L) \xrightarrow{\delta_L} H^2(M,G_t)_L \to$$
$$\to H^2(M,G_d)_R \xrightarrow{\rho_R{}^*} H^2(M,\mathfrak{m}^1_R)$$
$$\to H^2(M,G_d)_L \xrightarrow{\rho_L{}^*} H^2(M,\mathfrak{m}^1_L).$$

First six terms of these sequences were explicitly written by Oniscik ([16]). Before Oniščik, Röhrl used this sequence implicitly in the study of Riemann-Hilbert' problem ([20]). Vassiliou used a similar sequence derived from an operator algebra ([22]). Next three terms were known in the study of the second non abelian de Rham set ([4]). These are reviewed in Sections 3 and 4. The definition of the third non abelian de Rham set and the proof of the exactness of the above chiral sequences are given in Section 5.

In the study of de Rham correspondence of $H^1(M,\mathfrak{m}^1)$, we encounter the gauge transformation (of 1-forms), connection forms and curvature forms of G-bundles. Hence *the second non abelian de Rham theory relates to the geometry of fibre bundles and the (classical) gauge field theory*. This aspect also shows that the geometry of fibre bundles are dominated by the first order differential operator d and its (1-form) connections in the sense of [3] (cf.[1],[5]). In other words, we may say that *the geometry of fibre bundles is the geometry of* 1-*forms*. In the study of de Rham correspondence of $H^2(M,\mathfrak{m}^1)$, we encounter the gauge transformation of 2-forms and 2-form connections of some second order differential operator L (determined relative to a gauge and its connection) in the sense of [3] (cf.[11],[24]). Hence we may say *the third non abelian de Rham sets relate to the geometry of* 2-*forms*. The explicit form of L is not simple (Section 7, cf.[14], [15]). The study of L suggests that there exist some relations between the third non abelian de Rham set (determined absolutely by the space) and some BRS cohomology type abelian cohomologies (determined relative to a gauge and its connection) (Section 8, cf.[26]). From the study of L, we also speculate that there exists an algebra (possibly nilpotent) over the ring of matrix valued differential forms and a differential operator acting on this algebra. Properties of L will be understood through this operator (Section 7, cf.[25]).

Recent developments of the string field theory opens new perspective of geometry ([9],[14],[15],[23],[25]). Atempts have been made to understand the string field theory by the geometry (of 1-forms) of loop spaces or by the geometry of 2-forms ([9],[23]). Since the geometry of 1-forms over the loop space over M is transferred to the geometry of 2-forms over M, these two aspects are equivalent and relate to the third non abelian de Rham theory. In fact, we can regard the third non abelian de Rham theory to be the loop gauge theory (gauge theory over the loop space) or the theory of complete integrable systems over the double loop space Map (S^2,M). These aspects also suggest the possibility of the definitions of $H^p(M,\mathfrak{m}^1)$, $p\geqq 3$, by using complete integrable systems over interated loop spaces. Treaties on these subjects will appear soon (cf.[27]).

Similarly as abelian de Rham groups, non abelian de Rham sets are the obstruction sets of global solvabilities of *non linear* equations over M. These equations (for right handed systems) are given in the following table. Relations among non abelian de Rham sets, geometry and physics are also given in the table.

	Equation	Geometry	Physics
$H^0(M,\mathfrak{m}^1)$	$dg=g\theta$ θ is given	Geometry of π_1 (M)	Complete integrable systems
$H^1(M,\mathfrak{m}^1)$	$d\theta+\theta\wedge\theta=\Theta$ Θ is given	Geometry of fibre bundles	(Classical) gauge theory
$H^2(M,\mathfrak{m}^2)$	$d\Phi+[\phi,\Phi]=\Psi$ $\Phi-(d\phi+\phi\wedge\phi)+\psi$ Ψ, ψ are given	Geometry of gauge with connections	? (may be string field theory)

In this table, Θ is given relative to a gauge and Ψ, ψ are given relative to a gauge and a prescribed connection ϕ. Geometry of fibre bundles is the *geometry of* 1-*forms* and geometry of gauge with connections is the *geometry of* 2-*forms*.

2. NON ABELIAN POINCARE LEMMA

Now we explain details. First we study a local problem that is, the problem of non abelian Poincaré lemma ([6]). For the sake of simplicity, we assume in this Section that $\mathfrak{m}^1=\mathfrak{m}^1{}_R$. Then the first non abelian de Rham set is the obstruction set for the global solvability of the equation

$$dg - g\theta = 0, \tag{1}$$

where θ is a global 1-form over M such that $d\theta+\theta\wedge\theta=0$. To solve (1) locally, we assume that θ is defined on a star-like neighborhood about the origin of R^n. On such a neihghborhood, d has the chain homotopy I, given by

$$I\phi_{i_1,\cdots,i_{p-1}}(x) = \int_0^1 t^p \sum_{j=1}^{n} x_j\phi_{j,i_1,\cdots,i_{p-1}}(xt)dt,$$

$$I\phi = \sum I\phi_{i_1,\cdots,i_{p-1}} dx_{i_1}\wedge\cdots\wedge dx_{i_{p-1}},$$

where $\phi=\sum\phi_{i_1,\cdots,i_p} dx_{i_1}\wedge\cdots\wedge dx_{i_p}$. Using this I, we set

$$I_\theta(\phi) = I(\phi\wedge\theta), \quad P_\theta(\phi)=(1-I_\theta)^{-1}(\phi) = \phi + \sum_{n=1}^{\infty} I_\theta^n(\phi).$$

Then, if ϕ is a p-form, we have

$$dP_\theta(\phi) = P_\theta(\phi)\wedge\theta+P_{-\theta}(d\phi-(-1)^p I(P_\theta(\phi)\wedge(d\theta+\theta\wedge\theta))). \tag{2}$$

Especially, taking $\phi=1$, i.e. the identity matrix valued constant function, we get

$$dP_\theta(1) = P_\theta(1)\wedge\theta + P_{-\theta}(I(P_\theta(1)\wedge(d\theta+\theta\wedge\theta))).$$

Hence to define a differential operator ρ_R by $\rho_R(g)=g^{-1}dg$, we obtain the following (local) chain homotopy for ρ_R.

$$\rho_R(P_\theta(1)) + P_\theta(1)^{-1}P_{-\theta}(I(P_\theta(1)\wedge(d\theta+\theta\wedge\theta))) = \theta. \tag{3}$$

Frobenius' theorem (the first non abelian Poincaré lemma) follows from (3).

Next we want to solve locally the equation

$$d\theta + \theta\wedge\theta = \Theta. \tag{4}$$

From now on, we use the following notations.

$$d^e\theta = d\theta + \theta\wedge\theta, \quad d^e_\phi\beta = d\beta + [\phi,\beta].$$

Here $[\alpha,\beta]=\alpha\wedge\beta-(-1)^{pq}\beta\wedge\alpha$, where $p=\deg\alpha$ and $q=\deg\beta$. We define an integral transformation J_ϕ relative to a 1-form ϕ by

$$J_\phi(\beta) = (-1)^{p-1}P_\phi(1)^{-1}P_{(-1)^{p-1}\phi}(I(P_\phi(1)\beta)), \quad P=\deg\beta.$$

We set $J_\phi\begin{pmatrix}\alpha\\ \beta\end{pmatrix} = \begin{pmatrix}J_\phi(\alpha)\\ J_\phi(\beta)\end{pmatrix}$ for a vector of differential forms $\begin{pmatrix}\alpha\\ \beta\end{pmatrix}$.

If $\deg\ \alpha=2$ and $\deg\ \beta=1$, we define a differential operator D^e by

$$D^e\begin{pmatrix}\alpha\\ \beta\end{pmatrix} = \begin{pmatrix}d^e{}_\beta\alpha\\ \alpha-d^e\beta\end{pmatrix} = \begin{pmatrix}d\alpha + [\beta,\alpha]\\ \alpha-(d\beta+\beta_\wedge\beta)\end{pmatrix}.$$

From the definition of D^e, $D^e\begin{pmatrix}\alpha\\ \beta\end{pmatrix} = \begin{pmatrix}0\\ 0\end{pmatrix}$ if and only if $\alpha=d^e\beta$.

Using J_β and D^e, we define a transformation T by

$$T\begin{pmatrix}\alpha\\ \beta\end{pmatrix} = \begin{pmatrix}\alpha\\ \beta\end{pmatrix} - J_\beta\left(D^e\begin{pmatrix}\alpha\\ \beta\end{pmatrix}\right). \tag{5}$$

By definition, $\alpha=d^e\beta$ if and only if $\begin{pmatrix}\alpha\\ \beta\end{pmatrix}$ is a fixed point of T.

Moreover, we can solve the equation (4) locally by the iteration of T. Precisely saying, let us set $T^n\begin{pmatrix}\Theta\\ \theta_0\end{pmatrix} = \begin{pmatrix}\Theta_n\\ \theta_n\end{pmatrix}$. If $\Theta_n=\Theta$ for all n, then $\lim_{n\to\infty}\theta_n=\theta$ exists locally, and we have $d^e\theta=\Theta$. This is *the second non abelian Poincaré lemma* given in [6]. We note that differential operators d^e and D^e are appeared in (3) and (5), respectively. The second non abelian de Rham set $H^1(M,\mathfrak{m}^1)$ and the third non abelian de Rham set $H^2(M,\mathfrak{m}^1)$ are the obstruction sets for the global solvabilities of the operators d^e and D^e, respectively.

3. THE FIRST AND THE SECOND NON ABELIAN DE RHAM SETS

The first non abelian de Rham set $H^0(M,\mathfrak{m}^1)$ is the set of global integrable connections over M. The maps δ_R: $H^0(M,\mathfrak{m}^1{}_R)\to H^1(M,G_t)$ and δ_L: $H^0(M,\mathfrak{m}^1{}_L)\to H^1(M,G_t)$ are defined as follows;

$$\delta_R(\theta)_{ij} = h_i h_j^{-1},\ \theta|U_i = \rho_R(h_i),\ \text{and}$$

$$\delta_L(\theta)_{ij} = h_i^{-1} h_j,\ \theta|U_i = \rho_L(h_i).$$

$\delta_R(\theta)$ and $\delta_L(\theta)$ give the monodromy representations of the equations $dg=g\theta$ and $dg=\theta g$, respectively. Riemann-Hilbert' problem asks under what conditions can a representation of $\pi_1(M)$ be realized as a monodromy representation of some differential equation of type $dg=g\theta$? Since $\delta(\theta)$ is trivial as a smooth bundle and the converse of this fact is also true, we have a solution of Riemann-Hilbert' problem in smooth category (cf.[16],[18],[20],[22]). For a $\theta \in H^0(M,\mathfrak{m}^1)$ we define a characteristic class $\beta^p(\theta)\ H^{2p-1}(M,C)$ by

$$\beta^p(\theta) = \left< \frac{(-1)^{p-1}(p-1)!}{(2\pi\sqrt{-1})^p(2p-1)!} \operatorname{Trace}(\overbrace{\theta\wedge\cdots\wedge\theta}^{2p-1}) \right>. \tag{6}$$

Here $<\phi>$ means the de Rham class of ϕ. If (1) has a global solution f on M, that is, $\theta=f^{-1}df$ on M, we have

$$\beta^p(\theta) = f^*(e_p).$$

Here e_p is the $(2p-1)$-th generator of $H^*(G,Z)$ ([5],[19]). Characteristic classes for the elements of $H^i(M,\mathfrak{m}^1)$, $i=1,2$, will be defined similarly in this Section and in Section 6.

Let $\mathfrak{U}=\{U_i\}$ be a locally finite open covering of M, and $\mathfrak{G}^1$ be the sheaf of germs of smooth matrix valued 1-forms over M. Then we define the coboundary map $\delta=\delta_R: C^1(\mathfrak{U},\mathfrak{m}^1{}_R) \to C^2(\mathfrak{U},\mathfrak{G}^1)$ by

$$\delta_R(\omega)_{ijk} = \omega_{jk} - \omega_{ik} + \omega_{ij}{}^{g_{jk}}, \tag{7$_R$}$$

$$\omega_{ij} = \rho_R(g_{ij}),\quad \beta^g = g^{-1}\beta g.$$

Similary, $\delta_L: C^1(\mathfrak{U},\mathfrak{m}^1{}_L) \to C^2(\mathfrak{U},\mathfrak{G}^1)$ is defined by

$$\delta_L(\omega)_{ijk} = {}^{g_{ij}}\omega_{jk} - \omega_{ik} + \omega_{ij}, \tag{7$_L$}$$

$$\omega_{ij} = \rho_L(g_{ij}),\quad {}^g\beta = g\beta g^{-1}.$$

Definitions of δ_R and δ_L depend on the choice of $\{g_{ij}\}$ such that $\rho(g_{ij})=\omega_{ij}$. $\delta_R(\omega)=0$ or $\delta_L(\omega)=0$ means $\delta_R(\omega)_{ijk}=0$ or $\delta_L(\omega)_{ijk}=0$, respectively, for some choice of $\{g_{ij}\}$. For this $\{g_{ij}\}$, we have

$$\delta_R(g)_{ijk} = g_{ij}g_{jk}g_{ki} = \text{constant, for any } i,j,k, \text{ or}$$

$$\delta_L(g)_{ijk} = g_{ki}g_{ij}g_{jk} = \text{constant, for any } i,j,k.$$

It is also shown if $\delta_R(g)_{ijk}$=constant, for any i,j,k, then $\delta_R(\rho_R(g))_{ijk}=0$ (and the same statement is valid for left handed system). Therefore, if $\{g_{ij}\}$ defines a G-bundle, then $\delta_R(\rho_R(g))_{ijk}$ $=\delta_L(\delta_L(g))_{ijk}=0$. Since $\delta_L(g)_{ijk}=\delta_R(g)_{kij}$, two conditions $\delta_R(g)_{ijk}=$ constant, for any i,j,k, and $\delta_L(g)_{ijk}=$ constant, for any i,j,k, are equivalent. Hence *there are no serious problem of chirality in the study of* $H^1(M,\mathfrak{m}^1)$.

Roughly speaking, the cohomologous relation in $Z^1(\mathfrak{U},\mathfrak{m}^1{}_R) = \{\omega\in C^1(\mathfrak{U},\mathfrak{m}^1{}_R)\,|\,\delta_R\omega=0\}$ is given by

$$\{\rho_R(g_{ij})\} \sim \{\rho_R(g'_{ij})\} \quad \text{if} \quad g'_{ij}=h_i g_{ij} h_j^{-1}.$$

Here $\{h_i\}\in C^0(\ ,G_d)$ and $d({}^{h_i}(\rho_R(g)_{ijk}))=0$ for any i,j,k. We define $H^1(\mathfrak{U},\mathfrak{m}^1)$ by this relation. The limit of $H^1(\mathfrak{U},\mathfrak{m}^1{}_R)$ with respect to the refinement of the covering of M is well defined. $H^1(M,\mathfrak{m}^1{}_R)$ is defined by this limit. $H^1(M,\mathfrak{m}^1{}_L)$ is similarly defined ([4]). By definitions, we can define the maps $\rho_R{}^*:H^1(M,G_d)\to H^1(M,\mathfrak{m}^1{}_R)$ and $\rho_L{}^*:H^1(M,G_d)\to H^1(M,\mathfrak{m}^1{}_L)$. Since $\rho(g)=0$ if and only if g is a constant map, ker ρ^* is the image of $H^1(M,G_t)$ in $H^1(M,G_d)$.

The de Rham correspondence of $H^1(M,\mathfrak{m}^1{}_R)$ is defined as follows; If $\delta_R(\omega)=0$, then there exists a colleciton of 1-forms $\{\theta_i\}\in C^1(\mathfrak{U},\mathfrak{m}^1)$ such that

$$\omega_{ij} = \theta_j - \theta_i^{\,g_{ij}}. \tag{8}$$

This is shown by using smooth partition of unity subordinate to $\mathfrak{U}$. Since $\omega_{ij}=\rho_R(g_{ij})$, (8) means

$$g_{ij}^{-1}dg_{ij} = \theta_j - \theta_i^{g_{ij}}. \qquad (8)'$$

Hence, if $\{g_{ij}\}$ defines a G-bundle ξ, then $\{\theta_i\}$ is a connection form of ξ. By this reason, we say $\{\theta_i\}$ *to be a connection (form) of* $\{\omega_{ij}\}$ (*or* $\{g_{ij}\}$, *where* $\omega_{ij}=\rho_R(g_{ij})$). The curvature (form) $\{\Theta_i\}$ of $\{\theta_i\}$ is defined by

$$\Theta_i = d^e\theta_i \quad (=d\theta_i+\theta_i\wedge\theta_i).$$

We have $\Theta_j=\Theta_i^{g_{ij}}$ for any i,j. Hence the Trace $(\overbrace{\Theta_i\wedge\cdots\wedge\Theta_i}^{p})$ is a global 2p-form over M. Since Θ_i satisfies the Bianchi identity $d^e_{\theta_i}\Theta_i=0$, this 2p-form is closed and we can show that its de Rham class is determined by the cohomology class of $\{\omega_{ij}\}$ in $H^1(M,\mathfrak{m}^1)$ and does not depend on the choice of connections of $\{\omega_{ij}\}$ ([4]). Hence we can define characteristic classes $Ch^p(<\omega>)\in H^{2p}(M,C)$ for $<\omega>\in H^1(M,\mathfrak{m}^1)$ by

$$Ch^p(<\omega>) = <(\frac{\sqrt{-1}}{2\pi})^p \frac{1}{p!}\mathrm{Trace}(\overbrace{\Theta\wedge\cdots\wedge\Theta}^{p})> . \qquad (9)$$

By definition, $Ch^p(\rho^*(<\xi>))$ is the (p-th) Chern character of ξ. Similarly, we can define Chern classes of $<\omega>$. But they are not integral classes in general ([4], cf.[21]). We note that although $\delta_R g_{ijk}$ may not be equal to 1 in general but the curvature Θ_i satisfies

$$\Theta_i^{\delta_R g_{ijk}} = \Theta_i, \quad \text{for any } i,j,k.$$

From the viewpoint of connections of differential operators ([1], [3],[5]), above process is interpreted as follows; We denote $\mathfrak{G}^p$ the sheaf of germs of matrix valued p-forms over M. Then if $\{\beta_i\}\in C^1(\mathfrak{U},\mathfrak{G}^p)$ satisfies $\beta_j=\beta_i^{g_{ij}}$ and $\beta_i^{\delta_R g_{ijk}}=\beta_i$, then $d^e_{\theta_i}(\beta_i)$ also satisfies that

$$d^e_{\theta_j}(\beta_j) = (d^e_{\theta_i}(\beta_i))^{g_{ij}} ,$$

if $\{\theta_i\}$ is a connection of $\{\omega_{ij}\}=\{\rho_R(g_{ij})\}$. If $\{\theta_i\}$ is a connection of $\{\omega_{ij}\}$, we have

$$d^e_{\theta_i}(d^e_{\theta_i}(\beta_i)) = [\Theta_i,\beta_i],$$

by denoting its curvature $\{d^e\theta_i\}$ by $\{\Theta_i\}$. In other words, *the above process is dominated by the first order differential operator* d. Therefore we may say that the second non abelian de Rham theory relates to the geometry of 1-forms.

By definition, $\mathfrak{G}^1$ is a subsheaf of $\mathfrak{m}^1$. We define the coboundary map $\delta_{\xi,R}:C^1(\mathfrak{U},\mathfrak{G}^1)\to C^2(\mathfrak{U},\mathfrak{G}^1)$ by

$$\delta_{\xi,R}(\phi)_{ijk} = \phi_{jk} - \phi_{ik} + \phi_{ij}{}^{g_{jk}}.$$

Here $\xi=\{g_{ij}\}\in C^1(\mathfrak{U},G_d)$ is assumed to satisfy

$$g_{ii}=1,\ g_{ij}=g_{ji}{}^{-1} \text{ and } \delta_R g_{ijk}=\text{constant}. \qquad (*)$$

We set $Z^1(\mathfrak{U},\mathfrak{G}^1)=\{\phi\in C^1(\mathfrak{U},\mathfrak{G}^1)|\ \delta_{\xi,R}\phi=0$ for some $\xi\}$. Then we can define $H^1(M,\mathfrak{G}^1)$ and there is the inclusion $H^1(M,\mathfrak{m}^1)\to H^1(M,\mathfrak{G}^1)$. Here $H^1(M,\mathfrak{G}^1)$ is equal to 0 and the existence of connections for the elements $H^1(M,\mathfrak{m}^1)$ follows from the vanishing property of $H^1(M,\mathfrak{G}^1)$. If M is a complex manifold we can consider non abelian de Rham sets and their de Rham correspondences in holomorphic category. In this case, $H^1(M,\mathfrak{G}^1_\omega)$ may not be equal to 0. $\xi\in H^1(M,G_\omega)$ has a holomorphic connection if and only if the image of $\rho^*(\xi)$ in $H^1(M,\mathfrak{G}^1_\omega)$ is equal to 0 (cf.[8]). Hence G_ω and $\mathfrak{G}^1_\omega$ are the corresponding sheaves of G_d and $\mathfrak{G}^1$ in holomorphic category.

4. DEFINITIONS OF $H^2(M,G_*)$

The second cohomology set with coefficients in a sheaf of non abelian groups has been defined and studied by Dedecker since 1960 ([10], cf.[2]), and even a book by Giraud exists ([13]). Nevertheless the definition of $H^2(M,G_*)$ seems not so popular. So we sketch the definition of $H^2(M,G_*)$. Our definition of $H^2(M,G_*)$ differs slightly from Dedecker's definition. This is the reason why we obtain ten terms exact sequence stated in Introduction.

In the second non abelian cohomology, the coboundary map is defined relative to a chain $\xi=\{g_{ij}\}\in C^1(\mathfrak{U},G_*)$. We always assume that ξ satisfies the condition (*). We set

$$C^1(\mathfrak{U},G_*)_* = \{\xi\in C^1(\mathfrak{U},G_*)\,|\,\xi \text{ satisfies } (*)\}.$$

For a fixed ξ, there are two kinds of definitions $\delta_{\xi,R}$ and $\delta_{\xi,L}$ of coboundary maps. They are given by

$$\delta_{\xi,R}(f)_{ijk\ell} = f_{ij\ell}f_{jk\ell}(({f_{ijk}}^{-1})^{g_{k\ell}}){f_{ik\ell}}^{-1}, \tag{10$_R$}$$

$$\delta_{\xi,L}(f)_{ijk\ell} = {f_{ik\ell}}^{-1}{f_{ijk}}^{-1}({}^{g_{ij}}f_{jk\ell})f_{ij\ell}, \tag{11$_L$}$$

where $\{f_{ijk}\}$ belongs in $C^2(\mathfrak{U},G_*)$ (in [4], only $\delta_{\xi,L}$ was considered). From these definitions, we have

$$\delta_{\xi,R}(\delta_L\xi) = 1, \quad\text{and}\quad \delta_{\xi,L}(\delta_R\xi) = 1, \tag{11}$$

where $(\delta\xi)_{ijk}= g_{ijk}$. We call $\{f\}\in C^2(\mathfrak{U},G_*)$ a cocycle if $\delta_{\xi,R}f=1$ (or $\delta_{\xi,L}f=1$) for some ξ. Hence the cocycle condition depends on the domain of ξ. We assume that *the domain of* ξ *is* $C^1(\mathfrak{U},G_d)_*$ *although* f *is in* $C^2(\mathfrak{U},G_t)$. Precisely saying, we set

$$Z^2(\mathfrak{U},G_*)_R = \{f\in C^2(\mathfrak{U},G_*)\,|\,\delta_{\xi,R}f=1 \text{ for some } \xi\in C^1(\mathfrak{U},G_d)_*\},$$

$$Z^2(\mathfrak{U},G_*)_L = \{f\in C^2(\mathfrak{U},G_*)\,|\,\delta_{\xi,L}f=1 \text{ for some } \xi\in C^1(\mathfrak{U},G_d)_*\}.$$

Roughly speaking, $\{f\}, \{f'\}\in Z^2(\mathfrak{U},G_*)_R$ are said to be cohomologous if

$$f'_{ijk} = {h_{ik}}^{-1}f_{ijk}({h_{ij}}^{g_{jk}})h_{jk}, \quad \text{for some } \{h_{ij}\}\in C^1(\mathfrak{U},G_*),$$

where $\delta_{\xi,R}f=1$ and $\xi=\{g_{ij}\}$. Here $\{h_{ij}\}$ and $\{f'\}$ are assumed to satisfy

$$\xi'\in C^1(\mathfrak{U},G_d)_*,\quad \xi'=\{g'_{ij}\},\quad g'_{ij}=h_{ij}g_{ij}\quad\text{and}\quad \delta_{\xi',R}f'=1.$$

Similarly, under the assumptions

$$\xi'\in C^1(\mathfrak{U},G_d)_*,\ \xi'=\{g'_{ij}\},\ g'_{ij}=g_{ij}h_{ij},\ \{h_{ij}\}\in C^1(\mathfrak{U},G_*),$$

$$f'_{ijk}=h_{ij}({}^{g_{ij}}h_{jk})f_{ijk}h_{ik}{}^{-1},\ \xi=\{g_{ij}\},\ \delta_{\xi,L}f=\delta_{\xi',L}f'=1,$$

we say $\{f\}$ and $\{f'\}$ are cohomologous. Moreover, in the definitions of $H^2(\mathfrak{U},G_*)_R$ and $H^2(\mathfrak{U},G_*)_L$, we need to identify $\{f_{ijk}\}$ and $\{f_{ijk}{}^{h_k}\}$ or $\{f_{ijk}\}$ and $\{{}^{h_i}f_{ijk}\}$, $\{h_i\}\in C^1(\mathfrak{U},G_*)$. Then we can define cohomology sets $H^2(M,G_*)_R$ and $H^2(M,G_*)_L$. By (11), we can define maps $\delta_R:H^1(M,\mathfrak{m}^1{}_R)\to H^2(M,G_t)_L$ and $\delta_L:H^1(M,\mathfrak{m}^1{}_L)\to H^2(M,G_t)_R$ by by

$$\delta_R(\langle\omega\rangle) = \langle\{\delta_R(g)_{ijk}\}\rangle,\quad \omega_{ij}=\rho_R(g_{ij}),\quad\text{and}$$

$$\delta_L(\langle\omega\rangle) = \langle\{\delta_L(g)_{ijk}\}\rangle,\quad \omega_{ij}=\rho_L(g_{ij}).$$

In [4], only $H^2(M,G_*)_L$ was considered. The use of both $H^2(M,G_*)_R$ and $H^2(M,G_*)_L$ is necessary to get the ten terms of the exact sequence described in Introduction.

If $\{f_{ijk}\}$ is a representing cocycle of an element of $H^2(M,G_t)$ (or $H^2(M,G_d)$), then $\{\det f_{ijk}\}$ belongs to $Z^2(\mathfrak{U},C^*)$ (or $Z^2(\mathfrak{U},C^*{}_d)$). Its cohomology class in $H^2(M,C^*)$ (or in $H^2(M,C^*{}_d)\cong H^3(M,Z)$) is determined by the cohomology class of $\{f_{ijk}\}$ in $H^2(M,G_t)$ (or in $H^2(M,G_d)$). Hence we can define the maps

$$\det:\ H^2(M,G_t)\to H^2(M,C^*),$$
$$\det:\ H^2(M,G_d)\to H^2(M,C^*{}_d)\cong H^3(M,Z).$$

In the next Section, we define the third non abelian de Rham set

$H^2(M,\mathfrak{m}^1)$ and the map $\rho^*:H^2(M,G_d)\to H^2(M,\mathfrak{m}^1)$. Then the de Rham 3-form of $\det(\langle f\rangle)$, with $\langle f\rangle \in H^2(M,G_d)$, is realized by the de Rham 3-form of $\rho^*(f)\in H^2(M,\mathfrak{m}^1)$.

Under suitable assumptions, we can define higher characteristic classes $d^p(\langle c\rangle)\in H^{3p-1}(M,C^*)$, $\langle c\rangle\in H^2(M,G_t)$ and $e^p(\langle f\rangle)\in H^{3p-1}(M,C^*{}_d)\cong H^{3p}(M,Z)$ ([5],[7]). But details on these classes are not known.

5. RIGHT AND LEFT THIRD NON ABELIAN DE RHAM SETS

In this Section, we again assume that $\xi=\{g_{ij}\}$ belongs to $C^1(\mathfrak{U},G_d)_*$. If g is a smooth G-valued function and ϕ is a 1-form, we denote the right gauge transformation $\phi^g+\rho_R(g)$ by $g_R(\phi)$ and the left gauge transformation ${}^g\phi+\rho_L(g)$ by $g_L(\phi)$. We define the coboundary maps $\delta_{\xi,R}$: $C^2(\mathfrak{U},\mathfrak{m}^1{}_R)\to C^3(\mathfrak{U},\mathfrak{g}^1)$ and $\delta_{\xi,L}:C^2(\mathfrak{U},\mathfrak{m}^1{}_L)\to C^3(\mathfrak{U},\mathfrak{G}^1)$ by

$$\delta_{\xi,R}(\omega)_{ijk\ell} = \omega_{jk\ell} - ({g_{k\ell}}^{-1}{}_R(\omega)_{ik\ell})^{f_{ijk}g_{k\ell}} + \omega_{ij\ell}{}^{f_{jk\ell}} - g_{k\ell R}(\omega_{ijk}), \tag{12$_R$}$$

$$\delta_{\xi,L}(\omega)_{ijk\ell} = g_{ijL}(\omega_{jk\ell}) - {}^{f_{ijk}}\omega_{ik\ell} + {}^{g_{ij}f_{jk\ell}}({g_{ij}}^{-1}{}_L(\omega_{ij\ell})) - \omega_{ijk}. \tag{12$_L$}$$

Here, $\omega_{ijk}=\rho_R(f_{ijk})$ in $(12)_R$ and $\omega_{ijk}=\rho_L(f_{ijk})$ in $(12)_L$. The definition of δ_ξ depends on the choice of f such that $\rho(f)=\omega$. $\delta_\xi\omega=0$ means $\delta_\xi\omega_{ijk\ell}=0$ for some choice of $\{f_{ijk}\}$ such that $\rho(f_{ijk})=\omega_{ijk}$. For this $\{f_{ijk}\}$, we have

$$\delta_{\xi,R}(f)_{ijk\ell} = \text{constant}, \quad \text{if} \quad \delta_{\xi,R}(\omega)_{ijk\ell}=0, \tag{13$_R$}$$

and

$$\delta_{\xi,L}(f)_{ijk\ell} = \text{constant}, \quad \text{if} \quad \delta_{\xi,L}(\omega)_{ijk\ell}=0. \qquad (13)_L$$

The cocycle set $Z^2(\mathfrak{U},\mathfrak{m}^1{}_R)$ and $Z^2(\mathfrak{U},\mathfrak{m}^1{}_L)$ are given by

$$Z^2(\mathfrak{U},\mathfrak{m}^1{}_R) = \{\omega\in C^2(\mathfrak{U},\mathfrak{m}^1{}_R)|\delta_{\xi,R}\omega=0 \quad \text{for some} \quad \xi\in C^1(\mathfrak{U},G_d)_*\},$$

and

$$Z^2(\mathfrak{U},\mathfrak{m}^1{}_L) = \{\omega\in C^2(\mathfrak{U},\mathfrak{m}^1{}_L)|\delta_{\xi,L}\omega=0 \quad \text{for some} \quad \xi\in C^1(\mathfrak{U},G_d)_*\},$$

In $Z^2(\mathfrak{U},\mathfrak{m}^1)$, we define "strong" relation $\omega \sim \omega'$ by

$\omega=\rho_R(f)$, $\delta_{\xi,R}\omega=0$ by these $\{f\}$ and $\omega'=\rho_R(f')$, where

$$f'_{ijk} = c_{ijk}h_{ik}{}^{-1}f_{ijk}(h_{ij}{}^{g_{jk}})h_{jk},$$

for some $\{c_{ijk}\}\in C^2(\mathfrak{U},G_t)$ and $\{h_{ij}\}\in C^1(\mathfrak{U},G_d)$, or

$\omega=\rho_L(f)$, $\delta_{\xi,L}\omega=0$ by these $\{f\}$ and $\omega'=\rho_L(f')$, where

$$f'_{ijk} = h_{ij}({}^{g_{ij}}h_{jk})f_{ijk}h^{-1}_{ik}c_{ijk},$$

for some $\{c_{ijk}\}\in C^2(\mathfrak{U},G_t)$ and $\{h_{ij}\}\in C^1(\mathfrak{U},G_d)$. These strong relations are not equivalence relations. But we can generate equivalence relations from strong relations. We call them the cohomologous relations. We define $H^2(\mathfrak{U},\mathfrak{m}^1{}_R)$ and $H^2(\mathfrak{U},\mathfrak{m}^1{}_L)$ by cohomologous relations and the adjoint (or gauge) relations which are given by

$$\{\omega_{ijk}\} \sim \{((\omega_{ijk}+\rho_L(h_i)-\rho_L(h_i)^{f_{ijk}})^{h_i}\}, \quad \omega=\rho_R(f), \text{ or}$$

$$\{\omega_{ijk}\} \sim \{{}^{h_i}(\omega_{ijk}+\rho_R(h_i)-{}^{f_{ijk}}\rho_R(h_i))\}, \quad \omega=\rho_L(f).$$

The map $t^{\mathfrak{U}*}_{\mathfrak{V}}:H^2(\mathfrak{U},\mathfrak{m}^1)\to H^2(\mathfrak{V},\mathfrak{m}^1)$ cannot be defined on $H^2(\mathfrak{V},\mathfrak{m}^1)$. Here $\mathfrak{V}$ is a refinement of $\mathfrak{U}$. To define $t^{\mathfrak{U}*}_{\mathfrak{V}}$, we assume that $\omega\in Z^2(\mathfrak{U},\mathfrak{m}^1)$ satisfies the following conditions:

$$\omega_{isj}^{\delta_{\xi,R}(f)_{ituj}}=\omega_{isj}, \quad \text{for any } i,s,t,u,j, \quad \omega=\rho_R(f), \quad \text{or}$$

$$^{\delta_{\xi,L}(f)_{ituj}}\omega_{isj}=\omega_{isj}, \quad \text{for any } i,s,t,u,j, \quad \omega=\rho_L(f).$$

We note that if $\{f\}\in Z^2(\mathfrak{U},G_d)$, then this condition is satisfied. Under these assumptions, we can define cohomology sets $H^2(M,\mathfrak{m}^1{}_R)$ and $H^2(M,\mathfrak{m}^1{}_L)$. By definitions and (13), we can define the maps $\rho_R{}^*:H^2(M,G_d)_R\to H^2(M,\mathfrak{m}^1{}_R)$ and $\rho_L{}^*:H^2(M,G_d)_L\to H^2(M,\mathfrak{m}^1{}_L)$ by

$$\rho_R{}^*(<\{f_{ijk}\}>) = <\{\rho_R(f_{ijk})\}>, \quad \text{and}$$

$$\rho_L{}^*(<\{f_{ijk}\}>) = <\{\rho_L(f_{ijk})\}>.$$

Since $\rho(f)=0$ if and only if f is a constant map, the kernel of ρ^* is the image of $H^2(M,G_t)$ in $H^2(M,G_d)$. Hence we obtain the ten terms of the chiral exact sequence described in Introduction.

To get the de Rham correspondence of the third non abelian de Rham set, we first express $H^2(M,\mathfrak{m}^1)$ by 2-forms. We call this *the first process*. Then by using this 2-form expression, we get the 3-form expression, i.e. the de Rham correspondence of $H^2(M,\mathfrak{m}^1)$. We call this *the second process*. They are done as follows; The first process: we call $\{\theta_{ij}\}\in C^1(\mathfrak{U},\mathfrak{G}^1)$ a *primary connection* of $\{\omega_{ijk}\}\in Z^2(\mathfrak{U},\mathfrak{m}^1)$ if it satisfies

$$\omega_{ijk} = \theta_{jk} - \theta_{ik}{}^{f_{ijk}} + \theta_{ij}{}^{g_{jk}}, \quad \delta_{\xi,R}\omega=0, \ \omega=\rho_R(f), \qquad (14)_R$$

$$\omega_{ijk} = {}^{g_{ij}}\theta_{jk} - {}^{f_{ijk}}\theta_{ik} + \theta_{ij}, \quad \delta_{\xi,L}\omega=0, \ \omega=\rho_L(f). \qquad (14)_L$$

Here $\xi=\{g_{ij}\}$. The curvature form $\{\Theta_{ij}\}$ or $\{\theta_{ij}\}$ is defined by

$$\Theta_{ij} = d^e\theta_{ij}, \quad \text{if} \quad \delta_{\xi,R}\omega=0, \quad \text{and}$$

$$\Theta_{ij} = d^{e,L}\theta_{ij} \ (=d\theta_{ij}-\theta_{ij}\wedge\theta_{ij}), \quad \text{if} \quad \delta_{\xi,L}\omega=0.$$

We call $\{\Theta_{ij}\}$ a *primary curvature* of $\{\omega_{ijk}\}$. The correspondence $\{\omega_{ijk}\}\to\{\Theta_{ij}\}$ is the first process. This correspondence is not defined on $H^2(M,\mathfrak{m}^1)$. $\langle\omega\rangle$ has a primary connection if it satisfies

$$(g_{k\ell}{}^{-1}{}_R(\omega_{ik\ell}))^{f_{ijk}} = g_{k\ell}{}^{-1}{}_R(\omega_{ik\ell}), \quad \delta_{\xi,R}\omega=0, \qquad (15)_R$$

$${}^{f_{jk\ell}}(g_{ij}{}^{-1}{}_L(\omega_{ij\ell})) = g_{ij}{}^{-1}{}_L(\omega_{ij\ell}), \quad \delta_{\xi,L}\omega=0. \qquad (15)_L$$

In fact, under the assumption (15), (12) becomes to

$$\delta_{\xi,R}(\omega)_{ijk\ell} = \omega_{jk\ell} - \omega_{ik\ell} + \omega_{ij\ell}{}^{f_{jk\ell}} - \omega_{ij\ell}{}^{g_{k\ell}}, \quad \text{or}$$

$$\delta_{\xi,L}(\omega)_{ijk\ell} = {}^{g_{ij}}\omega_{jk\ell} - \omega_{ik\ell} + {}^{f_{jk\ell}}\omega_{ij\ell} - \omega_{ijk}.$$

Hence we can construct $\{\theta_{ij}\}$ by using a smooth partition of unity subordinate to $\mathfrak{U}$. Primary curvatures satisfy

$$\Theta_{jk}-\Theta_{ik}{}^{f_{ijk}}+\Theta_{ij}{}^{g_{jk}}+ [\theta_{jk}-\rho_R(g_{jk}), \theta_{ij}{}^{g_{jk}}] = 0, \qquad (16)_R$$

where $\{\Theta_{ij}\}$ is a primary curvature of $\{\omega\}\in Z^2(\mathfrak{U},\mathfrak{m}^1{}_R)$, or

$${}^{g_{ij}}\Theta_{jk}-{}^{f_{ijk}}\Theta_{ik}+\Theta_{ij}- [\theta_{ij}-\rho_L(g_{ij}), {}^{g_{ij}}\theta_{jk}] = 0, \qquad (16)_L$$

where $\{\Theta_{ij}\}$ is a primary curvature of $\{\omega\}\in Z^2(\mathfrak{U},\mathfrak{m}^1{}_L)$.

The second process: Let $\{\Phi_i\}$ be in $C^1(\mathfrak{U},\mathfrak{G}^2)$ and $\{\phi_i\}$ be in $C^1(\mathfrak{U},\mathfrak{G}^1)$. Then we call the pair $\{\Phi_i,\phi_i\}$ a *connection of* $\{\Theta_{ij}\}$ or a *secondary connection* of $\{\omega\}\in Z^2(\mathfrak{U},\mathfrak{m}^1)$ if

$$\Theta_{ij} = \Phi_j - \Phi_i{}^{g_{ij}} - [\theta_{ij}-\rho_R(g_{ij}),\phi_i{}^{g_{ij}}], \qquad (17)_R$$

where $\{\Theta_{ij}\}$ is a primary curvature of $\{\omega\}\in Z^2(\mathfrak{U},\mathfrak{m}^1{}_R)$, or

$$\Theta_{ij} = {}^{g_{ij}}\Phi_j - \Phi_i + [\theta_{ij}-\rho_L(g_{ij}), {}^{g_{ij}}\phi_j], \qquad (17)_L$$

where $\{\Theta_{ij}\}$ is a primary connection of $\{\omega\}\in Z^2(\mathfrak{U},\mathfrak{m}^1{}_L)$. The curvature forms $\{\Psi_i,\psi_i\}$ of $\{\Phi_i,\phi_i\}$ is defined by

$$\begin{pmatrix}\Psi_i\\ \psi_i\end{pmatrix} = D^e\begin{pmatrix}\Phi_i\\ \phi_i\end{pmatrix}, \quad \text{or} \quad \begin{pmatrix}\Psi_i\\ \psi_i\end{pmatrix} = D^{e,L}\begin{pmatrix}\Phi_i\\ \phi_i\end{pmatrix} = \begin{pmatrix}d\Phi_i-[\phi_i,\Phi_i]\\ \phi_i+(d\phi_i-\phi_i\wedge\phi_i)\end{pmatrix}.$$

We call $\{\Psi_i,\psi_i\}$ a *secondary curvature* of $\{\omega_{ijk}\}\cdot\{\Theta_{ij}\}$ has a connection (pair of forms) if it satisfies

$$\Theta_{ik}{}^{f_{ijk}} = \Theta_{ik}.$$

Moreover, if the primary connection satisfies $\theta_{ik}{}^{f_{ijk}}=\theta_{ik}$, then *we can take above* $\{\phi_i\}$ *to be a connection of the gauge* $\xi = \{g_{ij}\}$.

The relation between $\{\Psi_i,\psi_i\}$ and $\{\Psi_j,\psi_j\}$ is not simple. For example, we have

$$\Psi_j-\Psi_i{}^{g_{ij}} = [\theta_{ij}-\rho_R(g_{ij}),\psi_i{}^{g_{ij}}] - [\theta_{ij}-\phi_j+\phi_i{}^{g_{ij}},\phi_j],$$

in the right handed system. But this formula shows that $\{\text{Trace }\Psi_i\}$ defines a global 3-form over M. It is a closed form and *represents the cohomology class of* $\{\textit{Trace }\omega_{ijk}\}$ in $H^2(M,\mathfrak{D}^1)\cong H^3(M,C)$. Here, $\mathfrak{D}^1$ is the sheaf of germs of closed 1-forms over M. If $\langle\omega\rangle=\rho^*(\langle f\rangle)$, $\langle f\rangle\in H^2(M,G_d)$, $\{\textit{Trace }\Psi_i\}$ *represents* $i^*(\det\langle f\rangle)$ *in* $H^3(M,C)$, where $i^*:H^3(M,Z)\to H^3(M,C)$ is the map induced from the inclusion $i:Z\to C$. Hence it is an integral class.

6. COUPLING OF RIGHT HANDED SYSTEM AND LEFT HANDED SYSTEM

The above complicated relation between $\{\Psi_i,\psi_i\}$ and $\{\Psi_j,\psi_j\}$ is eliminated if we couple right handed system and left handed system. For this purpose, we need to assume $\{\omega_{ijk}\}\in Z^2(\mathfrak{U},\mathfrak{m}^1{}_R)$ satisfies

$$\delta_{\xi,R}(\omega)_{ijk\ell} = 0 \quad \text{and} \quad \delta_{\xi,L}({}^g\omega)_{ijk\ell} = 0, \tag{18}$$

where $\xi=\{g_{ij}\}$ and $({}^{g}\omega)_{ijk}={}^{g_{ik}}\omega_{ijk}$. We also assume that $\{\omega_{ijk}\}$ has a primary connection $\{\theta_{ij}\}$ such that

$$\omega_{ijk} = \theta_{jk} - \theta_{ik} + \theta_{ij}^{\;g_{jk}}, \quad \text{and} \quad \theta_{ij}^{\;\delta_R(g)_{ijk}}=\theta_{ij}. \tag{19}$$

Then $\{\omega_{ijk}\}$ has a secondary connection $\{\Phi_i,\phi_i\}$ such that $\{\phi_i\}$ is a connection of ξ. Moreover, $\{\omega_{ijk}\}$ and $\{\phi_i\}$ relate each other by the equation

$$d^e_{\phi_k}\omega_{ijk} \; (=d\omega_{ikj}+ [\phi_k,\omega_{ijk}]) = 0. \tag{20}$$

By (19), $\theta_{ij}={}^{g_{ij}}\theta_{ij}$ is a primary connection of $\{{}^{g_{ik}}\omega_{ijk}\}$. We also put $\eta_{ij}=\theta_{ij}-\rho_R(g_{ij})$. Then we get

$${}^{g_{ij}}\eta_{ij} = \theta_{ij} - \rho_L(g_{ij}) \quad (=\eta_{ij}).$$

Therefore, by (16), denoting secondary curvatures of $\{\omega\}$ and $\{{}^{g}\omega\}$ (determined by $\{\theta\}$ and $\{\theta\}$) by $\{\Psi_{i,R},\psi_{i,R}\}$ and $\{\Psi_{i,L},\psi_{i,L}\}$, respectively, we obtain

$$\begin{pmatrix}\Psi_{j,R}+\Psi_{j,L}\\ \psi_{j,R}+\psi_{j,L}\end{pmatrix} - \begin{pmatrix}\Psi_{i,R}+\Psi_{i,L}\\ \psi_{i,R}+\psi_{i,L}\end{pmatrix}^{g_{ij}} = 2\begin{pmatrix}[d^e\phi_j,\eta_{ij}]\\ d^e_{\phi_j}\eta_{ij}\end{pmatrix},$$

$$\Psi_{i,R} + \Psi_{i,L} = d^e_{\phi_i}(\psi_{i,R} + \psi_{i,L}).$$

Hence, if we define a 3-form Ψ_i and a 2-form ψ_i by

$$\Psi_i = \tfrac{1}{2}(\Psi_{i,R} + \Psi_{i,L}), \quad \psi_i = \tfrac{1}{2}(\psi_{i,R} + \psi_{i,L}),$$

then we have

$$\Psi_j - \Psi_i^{\;g_{ij}} = [d^e\phi_j,\eta_{ij}] \quad (=(d^e_{\phi_j})^2\eta_{ij}), \tag{21}$$

$$\psi_j - \psi_i^{\;g_{ij}} = d^e_{\phi_j}\eta_{ij},$$

and

$$\Psi_i = d^e_{\phi_i}\psi_i. \tag{22}$$

Therefore the correspondence $\{\omega_{ijk}\}\to\{\Psi_i\}$ is easier to treat than the de Rham correspondence $\{\omega_{ijk}\}\to\{\Psi_{i,R}\}$. So we call this correspondence the good de Rham correspondence. We call $\{\psi_i\}$ a 2-*form connection* of $\{\omega_{ijk}\}\in Z^2(\mathfrak{U},\mathfrak{m}^1_R)$ and $\{\Psi_i\}$ its 3-*form curvature* (cf.[23],[24]). We remark that if $\xi=\{g_{ij}\}$ does not define a fibre bundle, good de Rham correspondence is defined under the associativity condition (19) of adjoint actions by $\{g_{ij}\}$ (cf.[15],[23]).

Formally, we can eliminate the right hand side of the first formula of (21). That is, we can choose $\{\beta_i\}\in C^0(\mathfrak{U},\mathfrak{G}^2)$ such that

$$\Psi'_j = \Psi'_i{}^{g_{ij}}, \quad \Psi'_i = \Psi_i + d^e_{\phi_i}\beta_i.$$

In general, we cannot write this Ψ'_i in the form

$$\Psi'_i = \frac{1}{2}(\Psi_{i,R}+\Psi_{i,L}).$$

To define higher characteristic classes of $H^2(M,\mathfrak{m}^1)$ by using its 3-form curvature, we need to assume

$$\Psi_j = \Psi_i{}^{g_{ij}}, \quad \text{i.e. } [d^e\phi_j,\eta_{ij}] = 0 \text{ in (21).}$$

Even if this is satisfied, $\mathrm{Trace}(\overbrace{\Psi_i\wedge\cdots\wedge\Psi_i}^{p})$ may not be closed in general, if $p\geqq 2$. It becomes a global closed 3p-form over M for all p if Ψ_i satisfies

$$[d^e\phi_i,\Psi_i] \; (=(d^e_{\phi_i})^2\Psi_i) = 0.$$

On the other hand, we have

$$\langle\mathrm{Trace}(\overbrace{\Psi_i\wedge\cdots\wedge\Psi_i}^{p})\rangle = \langle\mathrm{Trace}(\overbrace{\Psi'_i\wedge\cdots\wedge\Psi'_i}^{p})\rangle,$$

for all p, if

$$d^e_{\phi_i}\Psi_i = d^e_{\phi_i}\Psi'_i = 0,$$

$$\Psi'_i = \Psi_i + d^e_{\phi_i}\beta_i, \quad \text{for some } \{\beta_i\} \text{ such that } \beta_j = \beta_i^{g_{ij}}.$$

These suggest us that higher characteristic classes are defined only on some special subsets of $H^2(M,\mathfrak{m}^1)$. If they are defined, then they take values in $H^{3+6p}(M,C)$, $p\geqq 0$. This also suggests that characteristic classes of $H^2(M,G_d)$ at $H^{6p}(M,Z)$, $p\geqq 1$, must be torsion classes if they exist (cf.[5],[7]).

It seems that there exist chiral maps

$$\ell_R: H^2(M,\mathfrak{m}^1_R) \to H^1(\Omega M,\mathfrak{m}^1_L),$$

$$\ell_L: H^2(M,\mathfrak{m}^1_L) \to H^1(\Omega M,\mathfrak{m}^1_R),$$

where ΩM is the loop space over M. If this is true, above discussions suggest us that $c^p(\ell(<\omega>)) \in H^{2p}(\Omega M,C)$ cannot be transgressive unless $p=3q+1$. Existence of ℓ also shows that the elements of $H^2(M,\mathfrak{m}^1)$ are expressed as flat connections over the double loop space $\Omega^2 M = \mathrm{Map}(S^2,M)$. Treaties on these subjects will appear soon (cf.[27]).

7. THIRD NON ABELIAN DE RHAM SETS AND GEOMETRY OF 2-FORMS

In this Section, a gauge ξ means an element $\xi=\{g_{ij}\}$ of $C^1(\mathfrak{U},G_d)_*$. The gauge transformation of a 1-form θ by ξ is defined by (8) or (8)'. Therefore we define the gauge transformation of a 2-form Φ by ξ in order to apply the second formula of (21). Since (21) contains $\{\theta_{ij}\}$, the gauge transformation of 2-forms by ξ is not well defined by ξ. To define the gauge transformation of ξ for 2-forms, in addition to the gauge $\xi=\{g_{ij}\}$, first we must assign a connection $\{\phi_i\}$ of ξ. Next we must specify $\{\theta_{ij}\} \in C^1(\mathfrak{U},\mathfrak{G}^1)$. Using these specifications, we define (right) gauge transformation: $(g_{ij},\phi_j,\theta_{ij})(\Phi)$ $(=g_{ij},\phi_j,\theta_{ij})_R(\Phi))$ of a 2-form Φ by

$$(g_{ij},\phi_j,\theta_{ij})(\Phi) = {}^{g_{ij}}\Phi + d^e_{\phi_j}(\theta_{ij} - \rho_R(g_{ij})). \tag{24}$$

We say the adjoint action of ξ on $\{\Phi_i\} \in C^1(\mathfrak{U},\mathfrak{G}^2)$ to be associative if we have

$$\Phi_i^{\delta_R(g)_{ijk}} = \Phi_i, \quad \text{for any } i,j,k. \tag{25)'}$$

Under the assumption (25)', (24) satisfies

$$(g_{jk},\phi_k,\theta_{jk})((g_{ij},\phi_j,\theta_{ij})(\Phi_i)) = (g_{ik},\phi_k,\theta_{ik})(\Phi_i), \tag{25}$$

if and only if

$$d^e_{\phi_k}(\delta_{\xi,R}(\theta)_{ijk}) = 0. \tag{26}$$

Here, $\delta_{\xi,R}(\theta)_{ijk}=\theta_{jk}-\theta_{ik}+\theta_{ij}^{g_{jk}}$. Because we have $\delta_{\xi,R}(\rho_R(g))_{ijk}=0$. In other words, $\{d^e_{\phi_k}(\delta_{\xi,R}(\theta)_{ijk})\} \in C^2(\mathfrak{U},\mathfrak{G}^2)$ gives *the obstruction class of associativity* (associativity anomaly) of the gauge transformation (24). If ξ defines a fibre bundle, then (25)' is always satisfied (cf.[16]). If $\{\theta_{ij}\}=\{0\}$, then (26) is satisfied. In this case, (24) becomes to

$$(g_{ij},\phi_j,0)(\Phi) = {}^{g_{ij}}\Phi-[\phi_j,\rho_R(g_{ij})] + \rho_R(g_{ij})\wedge\rho_R(g_{ij}). \tag{24)'}$$

The gauge transformation (24) contains the term

$$-d^e_{\phi_j}(\rho_R(g_{ij})) = -[\phi_j,\rho_R(g_{ij})] + \rho_R(g_{ij})\wedge\rho_R(g_{ij}).$$

This term is uniquely determined by the pair $\{\xi,\{\phi_i\}\}$ and does not depend on $\{\theta_{ij}\}$. For this term, we have

$$d^e_{\phi_j}d\beta-(d^e_{\phi_i}d)^{g_{ij}}\beta = -d^e_{\phi_j}(\rho_R(g_{ij}))\wedge\beta+\rho_R(g_{ij})\wedge d^e_{\phi_j}\beta.$$

Let us define a (local) differential operator L_i by

$$L_i\beta = d^e_{\phi_i}d\beta - \phi_i\wedge d^e_{\phi_i}\beta \tag{27}$$

(cf.[11]), then we obtain

$$L_j\beta - (L_i^{g_{ij}})\beta = -d^e_{\phi_j}(\rho_R(g_{ij}))\wedge\beta.$$

Therefore we can write

$$L_j - L_i^{g_{ij}} = -d^e_{\phi_j}(\rho_R(g_{ij})). \tag{28}$$

By (28), if $\{\Phi_i\}\in C^0(\mathfrak{U},\mathfrak{G}^2)$ satisfies

$$(g_{ij},\phi_j,\theta_{ij})(\Phi_i) = \Phi_j,$$

that is, if $\{\Phi_i\}\in C^0(\mathfrak{U},\mathfrak{G}^2)$ is a 2-form connection of $\{\omega_{ijk}\}=\{\delta_{\xi,R}(\theta)_{ijk}\}$, then we have

$$(L_j-\Phi_j) - (L_i-\Phi_i)^{g_{ij}} = -d^e_{\phi_j}(\theta_{ij}). \tag{29}$$

Especially, if $d^e_{\phi_j}(\theta_{ij})=0$, $\{L_i-\Phi_i\}$ defines a global differential operator. In this case, $\{-\Phi_i\}$ *is a connection of* $\{L_i\}$ in the sense of [1],[2],[3],[5] (cf.[24]).

We set $L_{\Phi_i}=L-\Phi_i$. Then by (29), we obtain

$$\begin{aligned}&(d^e_{\phi_j}L_{\Phi_j}-L_{\Phi_j}d^e_{\phi_j})\beta - (d^e_{\phi_i}L_{\Phi_i}-L_{\Phi_i}d^e_{\phi_i})^{g_{ij}}\beta\\ &= -[d^e\phi_j,\theta_{ij}]\wedge\beta.\end{aligned} \tag{30}$$

Hence if $\{\theta_{ij}\}$ satisfies

$$(d^e_{\phi_j})^2\theta_{ij}\ (=[d^e\phi_j,\theta_{ij}]) = 0,$$

the collection of differential operators

$$\{[d^e_{\phi_i},L_{\Phi_i}]\} = \{d^e_{\phi_i}L_{\Phi_i} - L_{\Phi_i}d^e_{\phi_i}\}$$

defines a global operator.

By definition, we have

$$L\beta = (-1)^p(d_{R,\phi}\beta)\wedge\phi-\phi\wedge\phi\wedge\beta = (-1)^p(d^e{}_\phi\beta)\wedge\phi-[\phi\wedge\phi,\beta]. \qquad (31)$$

Here $d_{R,\phi}\beta=d\beta+\phi\wedge\beta$ and $p=\deg\beta$. Hence $L\beta$ becomes to an inner derivation if $d^e{}_\phi\beta=0$ (cf.[15]). In general, L is not a (second order) derivation. It satisfies

$$L(\alpha\wedge\beta) = L\alpha\wedge\beta+\alpha\wedge L\beta-(-1)^p d^e{}_\phi\alpha\wedge[\phi,\beta], \quad p = \deg\alpha. \qquad (32)$$

The analogy of (32) and the addition formula of dilogarithmic function suggests that there may exist a matrix algebra (possibly nilpotent) over the ring of matrix valued differential forms and a "good" differential operator acting on this algebra. L, $d^e{}_\phi$ and the adjoint action by ϕ appear as matrix elements of this operator (cf.[25]).

The above discussion shows that the theory of 2-form connections are dominated by $\{\omega_{ijk}\}=\{\delta_{\xi,R}(\theta)_{ijk}\}\in C^2(\mathfrak{U},\mathfrak{G}^2)$. In the above discussion we did not use the property that $\{\omega_{ijk}\}\in Z^2(\mathfrak{U},\mathfrak{m}^1)$. If $\{\omega_{ijk}\}$ belongs to $Z^2(\mathfrak{U},\mathfrak{m}^1)$ and if $f=f_{ijk}$ satisfies $\rho_R(f)=\omega$, f must satisfy the following non linear equation (33) under the associativity assumption (25) of the gauge transformation $(g,\phi,\theta)(\Phi)$.

$$f^{-1}df\wedge f^{-1}df - [\phi,f^{-1}df] = 0. \qquad (33)$$

Moreover, there must exist the following relations between $\{f_{ijk}\}$ and $\{g_{ij}\}$:

$$\theta_{ik}{}^{f_{ijk}} = \theta_{ik}, \quad \text{for any } i,j,k,$$

and

$$(g_{k\ell}{}^{-1}{}_R(\omega_{ik\ell}))^{f_{ij\ell}} = g_{k\ell}{}^{-1}{}_R(\omega_{ik\ell}), \quad \text{for any } i,j,k,\ell.$$

Therefore, a 2-form connection only comes from a third non abelian de Rham class under the above strong conditions.

8. GEOMETRY OF 2-FORMS AND ABELIAN COHOMOLOGY

We can associate some abelian cohomology classes for a gauge transformation of 2-forms (2-form connection). This suggests that there may exist some abelian structure inside the third non abelian de Rham theory (cf.[26]).

Similarly to Section 7, $\xi=\{g_{ij}\}$ means a gauge and we fix a connection $\phi=\{\phi_i\}$ of ξ. We denote $\mathbb{D}^p{}_{(\xi,\phi)}$ the sheaf of germs of matrix valued p-forms β such that $d^e{}_\phi\beta=0$, over M. Here $\beta=\{\beta_i\}$ is assumed to satisfy $\beta_j=\beta_i{}^{g_{ij}}$. The abelian coboundary map $\delta_\xi:C^q(\mathfrak{U},\mathbb{D}^p{}_{(\xi,\phi)})\to C^{q+1}(\mathfrak{U},\mathbb{D}^p{}_{(\xi,\phi)})$ is defined by

$$\delta_\xi(\beta)_{i_0\cdots i_{q+1}} = \sum_{r=0}^{q}(-1)^r\beta_{i_0\cdots i_{r-1}i_{r+1}\cdots i_{q+1}} + (-1)^{q+1}\beta_{i_0\cdots i_q}{}^{g_{i_q i_{q+1}}}. \tag{34}$$

We set $Z^q(\mathfrak{U},\mathbb{D}^p{}_{(\xi,\phi)})=\{\beta\in C^q(\mathfrak{U},\mathbb{D}^p{}_{(\xi,\phi)})\,|\,\delta_\xi\beta=0\}$. Then, if $\{\theta_{ij}\}$ satisfies the associativity condition:

$$\theta_{ij}{}^{\delta_R(g)_{ijk}} = \theta_{ij}, \quad \text{for any } i,j,k,$$

then $\{\omega_{ijk}\}=\{\delta_{\xi,R}(\theta)_{ijk}\}$ belongs to $Z^2(\mathfrak{U},\mathbb{D}^1{}_{(\xi,\phi)})$ by (26). Hence $\{\omega_{ijk}\}$ defines an element of $H^2(M,\mathbb{D}^1{}_{(\xi,\phi)})$. We denote the 2-form connection derived from $\{\theta_{ij}\}$ by $\{\Phi_i\}$, and set $d^e{}_{\phi_i}\Phi_i=\Psi_i$. Then Trace Ψ_i defines a global closed 3-form over M. Its de Rham class represents the cohomology class of $\{\text{Trace }\omega_{ijk}\}\in H^2(M,\mathbb{D}^1)\cong H^3(M,C)$. We may define higher characteristic classes of $\{\omega_{ijk}\}$ by using Trace $\Psi_i\wedge\cdots\wedge\Psi_i$. The situation is the same as in Section 6.

There are two ways of definitions of higher characteristic classes of the element $\beta\in H^p(M,T^q{}_{(\xi,\phi)})$. Both of them take values in $H^{r(p+q)}(M,C)$ for $r\geq 2$ (if r=1, definition is always possible). The first way is to use the cup product $\{(\beta)^r{}_{i_0\cdots i_{rp}}\}$ and consider the de Rham class of $\{\mathrm{Trace}(\beta)^r\}$. Here, $(\beta)^r$ is defined by

$$(\beta)^r{}_{i_0\cdots i_{rp}} = \beta_{i_0\cdots i_p}\wedge\beta_{i_p\cdots i_{2p}}\wedge\cdots\wedge\beta_{i_{(r-1)p}\cdots i_{rp}}.$$

Another way is to express β by (p+q)-form (de Rham correspondence) and apply the same method as above. Owing to the existence of a gauge action in the definition of δ_ξ, $\delta_\xi(\beta^r)$ may not be equal to 0. Hence $\mathrm{Trace}(\beta^r)$ may not be a cocycle. Similarly, since we have no Poincaré lemma for the operator $d^e{}_\phi$, we cannot obtain de Rham map for $H^p(M,\mathfrak{D}^q{}_{(\xi,\phi)})$. To use the notations of Section 2, we set

$$d^\phi = (1+I_\phi)^{-1}d(1+I_\phi).$$

Then we have

$$d^\phi = d^e{}_\phi - (1+I_\phi)^{-1}I(d^e{}_\phi)^2.$$

Hence, if $(d^e{}_\phi)^2=0$, we have Poincaré lemma for $d^e{}_\phi$. This means ξ is essentially an abelian gauge by the theorem of Ambrose-Singer. The problem for a non abelian gauge seems still open.

Using cohomologies with coefficients in $\mathfrak{D}^2{}_{(\xi,\phi)}$, we can treat the obstruction of associativity (associativity anomaly) for a gauge transformation of 2-forms $(g_{ij},\phi_j,\theta_{ij})(\Phi_i)$, $\{\Phi_i\}\in C^0(\mathfrak{U},\mathfrak{G}^2)$, as follows; By the first equality of (21), we define the associated gauge action $(g_{ij},\phi_j,\theta_{ij})(\Psi_i)$ for $\{\Psi_i\}\in C^0(\mathfrak{U},\mathfrak{G}^3)$ by

$$(g_{ij},\phi_j,\theta_{ij})(\Psi_i) = \Psi_i{}^{g_{ij}}+[d^e\phi_j,\theta_{ij}-\rho_R(g_{ij})]. \tag{35}$$

We assume the adjoint action of ξ on $\{\Psi_i\}$ is associative. That is, we assume

$$\Psi_i{}^{\delta_R(g)_{ijk}} = \Psi_i, \quad \text{for any } i,j,k.$$

Then we have

$$(g_{jk},\phi_k,\theta_{jk})((g_{ij},\phi_j,\theta_{ij})(\Psi_i)) = (g_{ik},\phi_k,\theta_{ik})(\Psi_i), \tag{36}$$

if and only if

$$[d^e\phi_j, \delta_{\xi,R}(\theta)_{ijk}] = 0. \tag{37)'$$

Since (37)' means

$$d^e{}_{\phi_j}(d^e{}_{\phi_j}(\delta_{\xi,R}(\theta)_{ijk})) = 0, \tag{37}$$

We obtain

$$\{\omega_{ijk}\} = \{d^e{}_{\phi_j}(\delta_{\xi,R}(\theta)_{ijk})\} \in Z^2(\mathfrak{U},\mathbb{C}^2{}_{(\xi,\phi)}). \tag{38}$$

The cohomology class $<\{\omega_{ijk}\}>$ of $\{\omega_{ijk}\}$ in $H^2(M,\mathbb{C}^2{}_{(\xi,\phi)})$ is *the obstruction class of associativity* (associativity anomaly) of the gauge transformation of 2-forms $(g_{ij},\phi_j,\theta_{ij})(\Phi_i)$. $<\{\omega_{ijk}\}>$ is only determined under the associativity assumption (37) of the induced gauge transformation (35). By definition, $<\{\omega_{ijk}\}>$ is equal to 0 if $[d^e\phi_j,\theta_{ij}]=0$. $\{\mathrm{Trace}\ \omega_{ijk}\}$ defines an element of $H^4(M,C)$ and $<\{\omega_{ijk}\}>$ may have characteristic classes with values in $H^{4p}(M,C)$, $p\geqq 2$. Hence, we may evaluate associativity anomaly of the gauge transformation of 2-forms by cohomology classes in $H^{4p}(M,C)$, $p\geqq 1$.

The third non abelian de Rham set is determined absolutely by the space. While the above BRS type abelian cohomology derived from the operator $d^e{}_\phi$ is determined relative to a gauge and its connection. Although in the study of geometry of 2-forms, it may be more natural to use the above abelian cohomology than to use the third non abelian de Rham theory.

REFERENCES

[1] Andersson, S. I. : Vector bundle connections and lifting of partial differential operators, Lect. Notes in Math., 905, Springer, Berlin-New York, 119-132 (1982).

[2] Andersson, S. I. : Non-abelian Hodge theory via heat flow, Lect. Notes in Math., 1209, Springer, Berlin-New York, 8-36 (1986).

[3] Asada, A. : Connection of differential oeprators, J. Fac. Sci. Shinshu Univ., 13, 87-102 (1978).

[4] Asada, A. : Curvature forms with singularities and non integral characteristic classes, Lect. Notes in Math. 1139, Springer, Berlin-New York, 152-168 (1985).

[5] Asada, A. : Flat connections of differential operators and related characteristic classes, Proc. XIII Int. Conff. Diff. Geo. Methods in Theor. Phys., Shumen, 1984, World Sci., Singapore, 220-234 (1986).

[6] Asada, A. : Non abelian Poincaré lemma, Lect. Notes in Math., 1209, Springer, Berlin-New York, 37-65 (1986).

[7] Asada, A. : Non abelian de Rham theories, to appear in Proc. Int. Coll. Diff. Geo., Hajduszoboslõ, 1984.

[8] Atiyah, M. F. : Complex analytic connections in fibre bundles, Trans. Amer. Math. Soc., 85, 181-207 (1957).

[9] Bowick, M. J. and Rajeev, S. G. : String theory as the Kähler geometry of loop space, Phys. Rev. Lett., 58, 535-538 and 1158 (1987).

[10] Dedecker, P. : Sur la cohomologie non abélienne, I,II., Canad. J. Math., 12, 231-251 (1960), 15, 84-93 (1963).

[11] Evans, M. and Ovrut, A. B. : Prepotentials in superstring world sheet supergravity, Phys. Lett., B186, 134-140 (1987).

[12] Gaveau, B. : Intégrales harmoniques non-abéliennes, Bull. Sc. Math., 2e série 106, 113-169 (1982).

[13] Giraud, J., Cohomologie Non-abélienne, Springer, Berlin-New York, 1971.

[14] Horowitz, G. T., Lykken, J., Rohm, R. and Strominger, A.: Purely cubic action for string field theory, Phys. Rev. Lett., 57, 283-286 (1986).

[15] Horowitz, G. T. and Strominger, A. : Translations as inner derivations and associativity anomalites in open string field theory, Phys. Lett., B185, 45-51 (1987).

[16] Oniščik, A. L. : On the classification of fiber spaces, Sov. Math., Doklady, 2, 1561-1564 (1961).

[17] Oniščik, A. L. : Connections with zero curvature and de Rham theorem, Sov. Math., Doklady, 5, 1654-1657 (1964).

[18] Oniščik, A. L. : Some concepts and applications of non-abelian cohomology theory, Trans. Moscow Math. Soc., 17, 49-98 (1967).

[19] Pekonen, O. E. T. : Invariants secondaires de fibrés plats, C.R. Acad. Sc. Paris, 304, 13-14 (1987).

[20] Röhrl, H. : Das Riemann-Hilbertsche Problem der Theorie der Linearen Differentialgleichungen, Math. Ann., 133, 1-25 (1957).

[21] Shaposnik, F. A. and Solomin, J. E. : Gauge field singularities and non integer topological charge, J. Math. Phys., 20, 2110-2114 (1979).

[22] Vassiliou, E. : Sur les connexions plates d'un fibré banachiques, C.R. Acad. Sc. Paris, 295, 353-356 (1982).

[23] Vourdas, A. : Loop gauge theory and group cohomology, J. Math. Phys., 28, 584-591 (1987).

[24] Witten, E. : Non-abelian bosonization in two dimensions, Commun. Math. Phys., 92, 455-472 (1984).

[25] Witten, E. : Non-commutative geometry and string field theory, Nucl. Phys., B268, 253-294 (1986).

[26] Wu, Y. S. and Zee, A. : Abelian gauge structure inside non-abelian gauge theories, Nucl. Phys., B258, 157-178 (1985).

[27] Asada, A.: Differential geometry of loop spares, loop gauge theory, and non abelian de Rham theory, Proc. Symp. Topological Aspects of Modern Physics, Tokyo, 1987.

Proc. Prospects
of Math. Sci.
World Sci. Pub.
41-62, 1988

ON A DIOPHANTINE APPROXIMATION OF REAL NUMBERS FROM BELOW

SHUNJI ITO
Department of Mathematics
Tsuda College
Tsuda-Machi, Kodaira, Tokyo 187
Japan

0. INTRODUCTION

We see sometimes the following continued fraction expansion of a real number x, $0 < x < 1$ (see [5]):

$$x = \cfrac{1}{a_1 - \cfrac{1}{a_2 - \ddots - \cfrac{1}{a_n - \ddots}}} \tag{0.1}$$

where digits a_i $(i \geqq 1)$ are integers greater than 2. The importance of the expansion is derived from the following property (Proposition 1.1 in Section 1): For any irational x, $0 < x < 1$, if the pair of integer (q,p) such that $q > 0$ and p and q are coprime satisfies the inequality

$$0 < q(qx - p) < 1,$$

then there exists an integer n such that

$$(q, p) = (q_n, p_n),$$

where $$\frac{p_n}{q_n} := \cfrac{1}{a_1 - \cfrac{1}{a_2 - \ddots - \cfrac{1}{a_n}}}$$

(Remember the Legendre's theorem on a simple (regular) continued fraction expansion.) In other words, the expansion (0.1) proposes us the nice approximation of x *from below*.

In this paper, we first discuss several fundamental properties of the approximation of x by $\frac{p_n}{q_n}$ and its relations to the simple continued fraction expansion of x. Secondly, we discuss the metrical theory on this approximation. The main result is as follows (Theorem 5.1 in Section 5): For $0 < \lambda < 1$

$$\lim_{N\to\infty} \frac{\#\{(q,p) \mid 0 < q(qx-p) < \lambda,\ (q,p)=1,\ 0 < q \leqq N\}}{\log N} = \frac{6}{\pi^2} \cdot \lambda ,$$

for almost all x, $0 < x < 1$.
(Compare with the Erdős Theorem [1],[2],[3].)

Main tool on this discussion is the algorithm T (transformation) on [0,1] which induces the expansion (0.1):

$$Tx = \left\lceil \frac{1}{x} \right\rceil - \frac{1}{x} ,$$

where $\lceil x \rceil$ is the smallest integer greater than x. However, the transformation T has a σ-finite invariant measure μ with density

$$d\mu = \frac{dx}{1 - x} .$$

Therefore, it is difficult to discuss the problem of limit distribution on the basis of ergodic theorems. In Section 4, we give another algorithm (transformation) U such that

$$Ux = \begin{cases} Tx & \text{on } [0,2/3) , \\ T^{k-2}x & \text{on } [\frac{k-1}{k}, \frac{k}{k+1}) , k \geqq 3 , \end{cases}$$

which is the jump transformation of T and has a finite invariant measure ν with density

$$d\nu = \begin{cases} \frac{dx}{2\log 2(1-x)} & \text{on } [0,2/3) , \\ \frac{dx}{2\log 2(2-x)} & \text{on } [2/3,1) . \end{cases}$$

Using this transformation, we get several metrical theorems as an application of ergodic theory. This idea of constructing a jump transformation with finite invariant measure instead of the original one, can also be found in my paper [2]. The author hopes that the reader would compare this paper with my paper [2].

1. DEFINITION AND PROPERTIES

Let $X = [0,1]$ and the map T be defined on X by

$$Tx = \left\lceil \frac{1}{x} \right\rceil - \frac{1}{x} , \tag{1.1}$$

where $\lceil x \rceil$ is the smallest integer greater than x. For each $x \in X$, put an integer valued function

$$a(x) = \left\lceil \frac{1}{x} \right\rceil \tag{1.2}$$

and denote the k-th digit ($k \geqq 1$) of x by

$$a_k = a_k(x) := a(T^{k-1}x),$$

then the map T induces a kind of continued fraction expansion of x:

$$x = \cfrac{1}{a_1 - \cfrac{1}{a_2 - \ddots \cfrac{}{} \cfrac{1}{a_n - T^n x}}} . \tag{1.3}$$

For each x X, we prepare matrices

$$A_k = \begin{pmatrix} a_k & -1 \\ 1 & 0 \end{pmatrix} \tag{1.4}$$

and

$$\begin{pmatrix} q_n & -q_{n-1} \\ p_n & -p_{n-1} \end{pmatrix} = A_1 A_2 \cdots A_n \text{ , } n \geqq 1 \text{ ,} \tag{1.5}$$

where $\begin{pmatrix} q_0 & -q_{-1} \\ p_0 & -p_{-1} \end{pmatrix} = \begin{pmatrix} 1 & 0 \\ 0 & 1 \end{pmatrix}$,

then the following properties are obtained by induction:

<u>Property 1.1</u>. *For any irrational* $x \in X$, *we have*

1) $0 < q_n < q_{n+1}$,

2) $\dfrac{p_{n-1}}{q_{n-1}} < \dfrac{p_n}{q_n} < x$,

3) $x = \dfrac{p_n - T^n x \cdot p_{n-1}}{q_n - T^n x \cdot q_{n-1}}$,

4) $x - \dfrac{p_n}{q_n} = \dfrac{T^n x}{q_n(q_n - T^n x \cdot q_{n-1})}$.

Therefore we call a pair (X, T) the algorithm (or transformation) which induces the approximation of x *from below*, and call the expansion (1.3) as $n \to \infty$ B-continued fraction expansion of x, and denote it by

$$x = \langle 0 : a_1 a_2 a_3 \cdots a_n \cdots \rangle \text{ .} \tag{1.6}$$

The following Proposition characterizes the property of the algorithm.

<u>Proposition 1.1</u>. *The set of fractions* $\dfrac{p_n}{q_n}$, $n \geqq 1$ *consists with the principle and mediant convergents smaller than* x, *that is,*

$$\{ \frac{p_n}{q_n} : n\geq 1 \} = \{ \frac{k\hat{p}_{2m-1} + \hat{p}_{2m-2}}{k\hat{q}_{2m-1} + \hat{q}_{2m-2}} : 1\leq k\leq \hat{a}_{2m},\ m=1,2,\cdots \} ,$$

where $\frac{\hat{p}_m}{\hat{q}_m}$, $m \geq 1$ *is the* m-th *simple (regular) convergent and* $\hat{a}_m$ *is the* m-th *digit of a simple continued fraction expansion of* x. (The definition of a simple continued fraction expansion is seen in Section 2.)

Proof. It is sufficient to show that for any $A > 0$,

$$\min_{\substack{x>p/q \\ 0<q\leq A \\ (q,p)=1}} (x - \frac{p}{q}) = x - \frac{p_n}{q_n} ,$$

where n is chosen as $q_n \leq A < q_{n+1}$. (See Theorem 2.6 in [4].)

Assume that there exists a rational number $\frac{p}{q}$ satisfying

$$\frac{p_n}{q_n} < \frac{p}{q} < x \quad \text{and} \quad q \leq A .$$

By 1) and 2) of Property 1.1, there exists an integer $m(\geq n)$ such that

$$\frac{p_m}{q_m} \leq \frac{p}{q} < \frac{p_{m+1}}{q_{m+1}} .$$

Noting the determinant of the matrix $\begin{matrix} q_{m+1} & q_m \\ p_{m+1} & p_m \end{matrix}$ is equal to -1, we have the inequality

$$q \geq q_m + q_{m+1} \ (\geq q_{m+1}) .$$

This contradicts the assumption.

Remark 1.1. Another proof of the Proposition 1.1 will be given in Section 2.

Remark 1.2. The following statement is well known [4]: For a given irrational $x > 0$, if a simple fraction $\frac{p}{q}$, $q > 0$ satisfies the inequality

$$q|qx - p| < 1 ,$$

then $\frac{p}{q}$ is a mediant or a principle convergent of x. Combining the Proposition 1.1 and the above statement, we know the following fact: For a given irrational $x > 0$, if a simple fraction $\frac{p}{q}$, $q > 0$, satisfies the inequality

$$0 < q(qx - p) < 1 ,$$

then there exists an integer n such that

$$(q,p) = (q_n,p_n) .$$

2. ON THE RELATION TO THE SIMPLE CONTINUED FRACTION EXPANSION

The simple (or regular) continued fraction expansion, which will be called S-continued fraction expansion for the sake of simplicity, is induced by using the algorithm $(X,\hat{T})$ as follows: Let a map $\hat{T}$ on X be defined by

$$\hat{T}x = \frac{1}{x} - \left[\frac{1}{x}\right] , \tag{2.1}$$

and put $\hat{a}(x) = \left[\frac{1}{x}\right]$ and $\hat{a}_k = \hat{a}_k(x) = \hat{a}(\hat{T}^{k-1}x)$, then the map T induces a simple continued fraction expansion of x:

$$x = \cfrac{1}{\hat{a}_1 + \cfrac{1}{\hat{a}_2 + \ddots \cfrac{1}{\hat{a}_n + \hat{T}^n_x}}} , \tag{2.2}$$

and we denote it by

$$x = [0 : \hat{a}_1,\hat{a}_2,\cdots,\hat{a}_n,\cdots] . \tag{2.3}$$

For each modular transformation

$$z_1 = \frac{cz + b}{az + b} ,$$

we induce a 2×2 matrix:

$$\begin{pmatrix} a & b \\ c & d \end{pmatrix},$$

and call it a matrix representation of modular transformation. The expansions (1.2) and (2.2) can be regarded as the combinations of modular transformations

$$T^{k-1}x = \frac{1}{a_k - T^k x} \quad \text{and} \quad \hat{T}^{k-1}x = \frac{1}{\hat{a}_k + \hat{T}^k x}.$$

Therefore, we consider sometimes the products of matrices

$$\begin{pmatrix} a_1 & -1 \\ 1 & 0 \end{pmatrix}\begin{pmatrix} a_2 & -1 \\ 1 & 0 \end{pmatrix}\cdots\begin{pmatrix} a_n & -1 \\ 1 & 0 \end{pmatrix}$$

and

$$\begin{pmatrix} \hat{a}_1 & -1 \\ 1 & 0 \end{pmatrix}\begin{pmatrix} \hat{a}_2 & -1 \\ 1 & 0 \end{pmatrix}\cdots\begin{pmatrix} \hat{a}_n & -1 \\ 1 & 0 \end{pmatrix}$$

instead of B-expansion (1.2) and S-expansion (2.2) of x, respectively. The following fraction equations:

(1) $$\frac{1}{a + \frac{1}{b+x}} = \frac{1}{(a+1) - \frac{(b-1) + x}{b+x}},$$

(2) $$\frac{b-1+x}{b+x} = \cfrac{1}{2 - \cfrac{1}{2 - \cfrac{}{\ddots \, 2 - \cfrac{x}{1+x}}}},$$ (with $b-1$ twos)

and

(3) $$\frac{\frac{1}{c-x}}{1 + \frac{1}{c-x}} = \frac{1}{c+1-x},$$

provide us Sublemma for the products of matrices.

<u>Sublemma</u>.

1) $$\begin{pmatrix} a & 1 \\ 1 & 0 \end{pmatrix}\begin{pmatrix} b & 1 \\ 1 & 0 \end{pmatrix} = \begin{pmatrix} a+1 & -1 \\ 1 & 0 \end{pmatrix}\begin{pmatrix} b & 1 \\ b-1 & 1 \end{pmatrix},$$

2) $$\begin{pmatrix} b & 1 \\ b-1 & 1 \end{pmatrix} = \overbrace{\begin{pmatrix} 2 & -1 \\ 1 & 0 \end{pmatrix} \cdots \begin{pmatrix} 2 & -1 \\ 1 & 0 \end{pmatrix}}^{b-1} \begin{pmatrix} 1 & 1 \\ 1 & 0 \end{pmatrix},$$

and

3) $$\begin{pmatrix} 1 & 1 \\ 1 & 0 \end{pmatrix} \begin{pmatrix} c & -1 \\ 1 & 0 \end{pmatrix} = \begin{pmatrix} c+1 & -1 \\ 1 & 0 \end{pmatrix}.$$

Using Sublemma, we have

Theorem 2.1. *For each irrational* $x \in X$, *let*

$$x = [0; \hat{a}_1 \, \hat{a}_2 \cdots \hat{a}_n \cdots]$$

be a simple continued fraction expansion of x, *and*

$$x = \langle 0; a_1 \, a_2 \cdots a_n \cdots \rangle$$

be a B-*continued fraction expansion of the same* x, *then each sequence of digits has the following relation*:

$$(a_1, a_2, \cdots, a_n, \cdots) = (\hat{a}_1+1, 2, \overbrace{2, \cdots, 2}^{\hat{a}_2-1}, \hat{a}_3+2, \overbrace{2, \cdots, 2}^{\hat{a}_4-1}, \cdots ,$$
$$\cdots, a_{2m-1}+2, \underbrace{2, 2, \cdots, 2}_{a_{2m-1}}, \cdots)$$

where $\overbrace{2 \cdots 2}^{b}$ *means that the length of a block of consecutive* 2 *in the sequence* $(a_1, a_2, \cdots)$ *is equal to* b.

Proof. The product of matrices associated with a simple continued fraction expansion of x is decomposed by the following manners: For any $n > 0$,

$$\begin{pmatrix} \hat{a}_1 & 1 \\ 1 & 0 \end{pmatrix} \begin{pmatrix} \hat{a}_2 & 1 \\ 1 & 0 \end{pmatrix} \cdots \begin{pmatrix} \hat{a}_{2n-1} & 1 \\ 1 & 0 \end{pmatrix} \begin{pmatrix} \hat{a}_{2n} & 1 \\ 1 & 0 \end{pmatrix}$$

$$= \begin{pmatrix} \hat{a}_1+1 & -1 \\ 1 & 0 \end{pmatrix} \begin{pmatrix} \hat{a}_2 & 1 \\ \hat{a}_2-1 & 1 \end{pmatrix} \cdots \begin{pmatrix} \hat{a}_{2n-1}+1 & -1 \\ 1 & 0 \end{pmatrix} \begin{pmatrix} \hat{a}_{2n} & 1 \\ \hat{a}_{2n}-1 & 1 \end{pmatrix}$$

(by (1) of Sublemma).

$$= \begin{pmatrix} \hat{a}_1+1 & -1 \\ 1 & 0 \end{pmatrix} \overbrace{\begin{pmatrix} 2 & -1 \\ 1 & 0 \end{pmatrix} \cdots \begin{pmatrix} 2 & -1 \\ 1 & 0 \end{pmatrix}}^{\hat{a}_2-1} \begin{pmatrix} 1 & 1 \\ 0 & 1 \end{pmatrix} \cdots$$

$$\cdots \begin{pmatrix} \hat{a}_{2n}+1 & -1 \\ 1 & 0 \end{pmatrix} \overbrace{\begin{pmatrix} 2 & -1 \\ 1 & 0 \end{pmatrix} \cdots \begin{pmatrix} 2 & -1 \\ 1 & 0 \end{pmatrix}}^{\hat{a}_{2n}-1} \begin{pmatrix} 1 & 1 \\ 0 & 1 \end{pmatrix}$$

(by (2) of Sublemma).

$$\begin{pmatrix} \hat{a}+1 & -1 \\ 1 & 0 \end{pmatrix} \overbrace{\begin{pmatrix} 2 & -1 \\ 1 & 0 \end{pmatrix} \cdots \begin{pmatrix} 2 & -1 \\ 1 & 0 \end{pmatrix}}^{\hat{a}_2-1} \begin{pmatrix} \hat{a}_3+2 & -1 \\ 1 & 0 \end{pmatrix} \overbrace{\begin{pmatrix} 2 & -1 \\ 1 & 0 \end{pmatrix} \cdots \begin{pmatrix} 2 & -1 \\ 1 & 0 \end{pmatrix}}^{\hat{a}_4-1}$$

$$\cdots \begin{pmatrix} a_{2n}+2 & -1 \\ 1 & 0 \end{pmatrix} \overbrace{\begin{pmatrix} 2 & 1 \\ 1 & 0 \end{pmatrix} \cdots \begin{pmatrix} 2 & -1 \\ 1 & 0 \end{pmatrix}}^{\hat{a}_{2n}-1} \begin{pmatrix} 1 & 1 \\ 0 & 1 \end{pmatrix}$$

(by (3) of sublemma).

The final form as n tends to infinity corresponds to a B-continued fraction expansion $x = <0;\hat{a}_1+1, \overbrace{2,\cdots,2}^{\hat{a}_2-1},\cdots>$.

<u>Remark 2.1</u>. The idea of reforming the product of matrices in the proof affords us another proof of Proposition 1.1. In fact, for each $x = <0,a_1a_2\cdots>$ and n, there exists an integer m such that

$$\sum_{j=1}^{m} \hat{a}_{2j} \leq n < \sum_{j=1}^{m+1} \hat{a}_{2j} \ ,$$

and the following equality holds for each n

$$(a_1,\cdots,a_n) = (\hat{a}_1+1,\overbrace{2,\cdots 2}^{\hat{a}_2-1},\hat{a}_3+2,\overbrace{2,\cdots 2}^{\hat{a}_4-1},\cdots a_{2m+1}+2,\overbrace{2,2,\cdots 2}^{n-\sum_{j=1}^{m} a_{2j}-1}) \ . \quad (2.4)$$

In other words, the product of matrices is reformed as follows:

$$\begin{pmatrix} a_1 & -1 \\ 1 & 0 \end{pmatrix} \cdots \begin{pmatrix} a_n & -1 \\ 1 & 0 \end{pmatrix} = \begin{pmatrix} \hat{a}_1 & 1 \\ 1 & 0 \end{pmatrix} \cdots \begin{pmatrix} \hat{a}_{2m} & 1 \\ 1 & 0 \end{pmatrix} A \ , \qquad (2.5)$$

$$\text{where } A = \begin{cases} \begin{pmatrix} 1 & -1 \\ 0 & 1 \end{pmatrix} \left(= \begin{pmatrix} 1 & 1 \\ 0 & 1 \end{pmatrix}^{-1} \right) & , \text{ if } n = \sum_{k=1}^{m} \hat{a}_{2j} \ , \\ \begin{pmatrix} 1 & 1 \\ 0 & 1 \end{pmatrix} \begin{pmatrix} \hat{a}_{2m+1}+1 & -1 \\ 1 & 0 \end{pmatrix} \overbrace{\begin{pmatrix} 2 & -1 \\ 1 & 0 \end{pmatrix} \cdots \begin{pmatrix} 2 & -1 \\ 1 & 0 \end{pmatrix}}^{k-1} \ , & \text{if } n > \sum_{j=1}^{m} \hat{a}_{2j} \ , \end{cases}$$

and $k = n - \sum_{k=1}^{m} \hat{a}_k$.

Now, from the definition, we have

$$\begin{pmatrix} q_n \\ p_n \end{pmatrix} = \begin{pmatrix} a_1 & -1 \\ 1 & 0 \end{pmatrix} \cdots \begin{pmatrix} a_n & -1 \\ 1 & 0 \end{pmatrix} \begin{pmatrix} 1 \\ 0 \end{pmatrix}$$

$$= \begin{pmatrix} \hat{a}_1 & 1 \\ 1 & 0 \end{pmatrix} \cdots \begin{pmatrix} \hat{a}_{2m} & 1 \\ 1 & 0 \end{pmatrix} A \begin{pmatrix} 1 \\ 0 \end{pmatrix} \qquad \text{(by (2.5))}$$

$$= \begin{cases} \begin{pmatrix} \hat{q}_{2m} \\ \hat{p}_{2m} \end{pmatrix} & , \text{ if } n = \sum_{k=1}^{m} \hat{a}_{2j} \ , \\ k \begin{pmatrix} \hat{q}_{2m+1} \\ \hat{p}_{2m+1} \end{pmatrix} + \begin{pmatrix} \hat{q}_{2m} \\ \hat{p}_{2m} \end{pmatrix} & , \text{ if } n > \sum_{j=1}^{m} \hat{a}_{2j} \ , \end{cases}$$

where k satisfies the inquality $1 \leqq k \leqq \hat{a}_{2(m+1)} - 1$. This is nothing but Proposition 1.1.

3. ON THE NATURAL EXTENSION OF B-CONTINUED FRACTION TRANSFORMATION

In this Section, the ergodic property of the transformation T is discussed.

We construct a map $\overline{T}$ which will be called a natural extension of T. Let $\overline{X} = [0,1] \times [0,1]$ and the map $\overline{T}$ on $\overline{X}$ is defined by

$$\bar{T}(x,y) = (Tx, \frac{1}{a_1 - y}) , \tag{3.1}$$

then we see that the map $\bar{T}$ is bijective on $\bar{X}$. Define the measure $\bar{\mu}$ on $\bar{X}$ by

$$d\bar{\mu}(x,y) = \frac{dxdy}{(1-xy)^2} . \tag{3.2}$$

Then we have

Theorem 3.1. *The measure* $\bar{\mu}$ *is a* σ*-finite invariant measure with respect to the map* $\bar{T}$, *and the dynamical system* $(\bar{X}, \bar{T}, \bar{\mu})$ *is ergodic.*

Taking the marginal distribution of $\bar{\mu}$, we have

Corollary 3.2. *Let* $d\mu = \frac{dx}{1-x} = \int_0^1 \frac{dy}{(1-xy)^2} dx$,

then the measure μ *is a* σ*-finite invariant measure with respect to* T, *and the dynamical system* (X, T, μ) *is ergodic.*

The introduction of the natural extension provides us the following lemma which gives an interpretation of diophantine approximation in terms of ergodic theory.

Lemma 3.1. *For any* $\lambda > 0$, (q_n, p_n) *satisfies the inequality*

$$q_n(q_n x - p_n) < \lambda$$

if and only if the value of n-th *iteration* $\bar{T}^n$ *of the point* $(x,0)$ *satisfies*

$$\bar{T}^n(x,0) \in D_\lambda ,$$

where $D_\lambda = \{(x,y) \in \bar{X} : \frac{x}{1-xy} < \lambda \}$.

Proof. By (4) of Property 1.1, we know

$$q_n(q_n x - p_n) = \frac{T^n x}{(1 - \frac{q_{n-1}}{q_n} \cdot T^n x)} . \tag{3.3}$$

By the way, from the definition of T, we have

$$\overline{T}^n(x,0) = (T^n x, \frac{q_{n-1}}{q_n}) . \tag{3.4}$$

Therefore, we get the conclusion.

Corollary 3.1. *For almost all* $x \in X$,

(1) $\inf_n q_n(q_n x - p_n) = 0$,

(2) $\sup_n q_n(q_n x - p_n) = \infty$.

Proof. Let λ_k be a sequence converging to zero. By the ergodicity of $(\overline{X}, \overline{T}, \overline{\mu})$, the points $\overline{T}^n(x,y)$ visit infinitly many often on the set D_{λ_k} for almost all (x,y). Remarking

$$d(\overline{T}^n(x,y), \overline{T}^n(x,0)) < \frac{1}{n} ,$$

and using Lemma 3.1, we obtain the conclusion (1). The assersion (2) is obtained similarly by taking $\lambda_k \to \infty$ and considering $D^c_{\lambda_k}$.

4. NEAREST MEDIANT CONVERGENT FROM BELOW

In this Section we propose a new algorithm which induces the principal and the nearest mediant convergent from below of x.

Let $$I_k = [\frac{1}{k}, \frac{1}{k-1}) , \ k \geq 3 ,$$

and $$I_{2,k} = [\frac{k-1}{k}, \frac{k}{k+1}) , \ k \geq 2 ,$$

and define the map U on $X = [0,1] = \bigcup_{k\geq 3} I_k \cup (\bigcup_{j\geq 2} I_{2,j})$

as a jump transformation of T by

$$Ux = \begin{cases} Tx & \text{on } I_k \text{ or } I_{2,2}, \\ T^{k-2}x & \text{on } I_{2,k},\ k \geqq 3, \end{cases} \tag{4.1}$$

$$= \begin{cases} k - \frac{1}{x} & \text{on } I_k,\ k \geqq 3, \\ 2 - \frac{1}{x} & \text{on } I_{2,2}, \\ \frac{-(k-2) + (k-1)x}{-(k-3) + (k-2)x} & \text{on } I_{2,k},\ k \geqq 3. \end{cases} \tag{4.2}$$

Corresponding to the inverse branches of U, we induce the matrices:

$$B_k = \begin{pmatrix} k & -1 \\ 1 & 0 \end{pmatrix}, \qquad B_{2,2} = \begin{pmatrix} 2 & -1 \\ 1 & 0 \end{pmatrix}, \tag{4.3}$$

and

$$B_{2,k} = \underbrace{\begin{pmatrix} 2 & -1 \\ 1 & 0 \end{pmatrix} \cdots \begin{pmatrix} 2 & -1 \\ 1 & 0 \end{pmatrix}}_{k-2} = \begin{pmatrix} k-1 & -(k-2) \\ k-2 & -(k-3) \end{pmatrix}.$$

Put

$$\delta_n = \delta_n(x) = \begin{cases} k & , \text{ if } U^{n-1}x \in I_k, \\ (2,k) & , \text{ if } U^{n-1}x \in I_{2,k}, \end{cases} \tag{4.4}$$

then digits δ_n have the following Markov properties:

(1) If δ_n is $(2,k)$ type and $k > 2$, then $\delta_{n+1} = (2,2)$.

(2) If δ_n is $(2,2)$ type, then δ_{n+1} is R-type.

Using the sequence of digits corresponding to each $x \in X$, we define the matrix:

$$\begin{pmatrix} r_n & s_n \\ t_n & u_n \end{pmatrix} = B_{\delta_1} B_{\delta_2} \cdots B_{\delta_n} \quad (n \geqq 1). \tag{4.5}$$

Then we have the following representation:

Proposition 4.1. *For any irrational* $x \in X$,

$$x = \frac{t_n + u_n \cdot U^n x}{I_n + S_n \cdot U^n x} . \tag{4.6}$$

From the definitions (4.3) and (4.5), we see the sequence of fractions $\frac{t_n}{r_n}$ is a subsequence of $\{ \frac{P_n}{q_n} : n \geqq 1\}$. The following theorem characterizes the fractions $\frac{t_n}{I_n}$.

Theorem 4.1. *The algorithm* U *induces the principle and nearest mediant convergents of* x *from below, that is,*

$$\left\{\frac{t_n}{r_n};\ n=1,2,\cdots\right\}=\left\{\frac{\hat{p}_{2m+1}+\hat{p}_{2m}}{\hat{q}_{2m+1}+\hat{q}_{2m}},\ \frac{(\hat{a}_{2(m+1)}-1)\ \hat{p}_{2m+1}+\hat{p}_{2m}}{(\hat{a}_{2(m+1)}-1)\ \hat{q}_{2m+1}+\hat{q}_{2m}},\ \frac{\hat{q}_{2(m+1)}}{\hat{q}_{2(m+1)}}\ ;\ m=0,1,2,\cdots\right\}.$$

Proof. If there exists in the formula (4.5) the matrix B_{δ_j} such that δ_j is $(2,k)$ $(k \geqq 3)$ type, replace B_{δ_j} with $\overbrace{\begin{pmatrix} 2 & -1 \\ 1 & 0 \end{pmatrix} \cdots \begin{pmatrix} 2 & -1 \\ 1 & 0 \end{pmatrix}}^{j-2}$. The product form of replaced matrices is nothing but the product of matrices associated with B-continued fraction expansion. In other word, the necessary and sufficient condition that the matrix associated with B-expansion:

$$\begin{pmatrix} q_n & -q_{n-1} \\ p_n & -p_{n-1} \end{pmatrix} = \begin{pmatrix} a_1 & -1 \\ 1 & 0 \end{pmatrix} \cdots \begin{pmatrix} a_n & -1 \\ 1 & 0 \end{pmatrix}$$

coincides with the matrix (associated with U-expansion):

$$\begin{pmatrix} s_\ell & s_\ell \\ t_\ell & t_\ell \end{pmatrix} = B_{\delta_1}\ B_{\delta_2} \cdots B_{\delta_\ell}$$

is that the value of index $n - \sum_{j=1}^{m} \hat{a}_{2j}$ is equal to 0, $\hat{a}_{2(m+1)}-2$ or $\hat{a}_{2(m+1)}-1$ on the representation (2.4):

$$(a_1,a_2,\cdots a_n) = (\hat{a}_1+1,\overbrace{2,\cdots 2}^{\hat{a}_2-1},\cdots,\hat{a}_{2m+1}+2,\overbrace{2,2,\cdots 2}^{n-\sum_{j=1}^{m}\hat{a}_{2j}}) .$$

Therefore, in the case of $n = \sum_{j=1}^{m}\hat{a}_{2j}$, $\sum_{j=1}^{m+1}\hat{a}_{2j}-2$ or $\sum_{j=1}^{m+1}\hat{a}_{2j}-1$, we have

$$\begin{pmatrix} q_n \\ p_n \end{pmatrix} = \begin{pmatrix} q_n - q_{n-1} \\ p_n - p_{n-1} \end{pmatrix}\begin{pmatrix} 1 \\ 0 \end{pmatrix}$$

$$= \begin{pmatrix} q_{2m} & q_{2m-1} \\ p_{2m} & p_{2m-1} \end{pmatrix} A' \begin{pmatrix} 1 \\ 0 \end{pmatrix},$$

where $A' = \begin{cases} \begin{pmatrix} \hat{a}_{2m+1}+2 & -1 \\ 1 & 0 \end{pmatrix} , \\ \begin{pmatrix} \hat{a}_{2m+1}+2 & -1 \\ 1 & 0 \end{pmatrix} \overbrace{\begin{pmatrix} 2 & -1 \\ 1 & 0 \end{pmatrix} \cdots \begin{pmatrix} 2 & -1 \\ 1 & 0 \end{pmatrix}}^{\hat{a}_{2(m+1)}-2} \\ \begin{pmatrix} \hat{a}_{2m+1}+2 & -1 \\ 1 & 0 \end{pmatrix} \underbrace{\begin{pmatrix} 2 & -1 \\ 1 & 0 \end{pmatrix} \cdots \begin{pmatrix} 2 & -1 \\ 1 & 0 \end{pmatrix}}_{\hat{a}_{2(m+1)}-1} \end{cases}$

respectively, that is, in the case of $n = \sum_{j=1}^{m}\hat{a}_{2j}$, $\sum_{j=1}^{m+1}\hat{a}_{2j}-2$ or $\sum_{j=1}^{m}\hat{a}_{2j}-1$, we have

$$\begin{pmatrix} q_n \\ p_n \end{pmatrix} = \begin{pmatrix} \hat{q}_{2m+1} \\ \hat{p}_{2m+1} \end{pmatrix} + \begin{pmatrix} \hat{q}_{2m} \\ \hat{p}_{2m} \end{pmatrix},\ (\hat{a}_{2(m+1)}-1)\begin{pmatrix} \hat{q}_{2m+1} \\ \hat{p}_{2m+1} \end{pmatrix} + \begin{pmatrix} \hat{q}_{2m} \\ \hat{p}_{2m} \end{pmatrix} \text{ or } \begin{pmatrix} \hat{q}_{2(m+1)} \\ \hat{p}_{2(m+1)} \end{pmatrix}.$$

Proposition 4.2. *For each irrational* $x \in X$,

(1) $$x - \frac{t_n}{I_n} = \frac{U^n x}{I_n^2(1 + \frac{S_n}{I_n} \cdot U^n x)},$$

$$(2)\quad r\,x - t_n = g(x)\,g(Ux)\cdots g(U^{n-1}x)\cdot U^n x$$

where $g(x) = \begin{cases} x & ,\ \text{if}\ \ x\ \ [0,2/3) \\ -(k-3)-(k-2)x & ,\ \text{if}\ \ x\ \ I_{2,k}\ \ (k \geqq 3)\ . \end{cases}$

<u>Proof</u>. The formula (1) is obtained from Proposition 4.1. For the formula (2), we consider an affine transformation φ_{δ_1} from $\begin{pmatrix}\xi_{i-1}\\ \eta_{i-1}\end{pmatrix}$-plane to $\begin{pmatrix}\xi_i\\ \eta_i\end{pmatrix}$-plane defined by

$$\varphi_{\delta_i}:\ \begin{pmatrix}\xi_{i-1}\\ \eta_{i-1}\end{pmatrix} = B_{\delta_i}\begin{pmatrix}\xi_i\\ \eta_i\end{pmatrix}\ .$$

The linear form $x\cdot\xi_0 - \eta_0$ is transformed by φ_{δ_1} as follows:

$$x\xi_0-\eta_0 = g(x)(Ux\cdot\xi_1-\eta_1)\ .$$

Therefore, we have inductively

$$x\xi_0-\eta_0 = g(x)g(Ux)\cdots g(U^{n-1}x)(U^n x\cdot\xi_n-\eta_n)\ ,\ n \geqq 1\ . \tag{4.7}$$

On the other hand, we know

$$\begin{pmatrix}\xi_0\\ \eta_0\end{pmatrix} = \begin{pmatrix} r_n & s_n \\ t_n & u_n\end{pmatrix}\begin{pmatrix}\xi_n\\ \eta_n\end{pmatrix}.$$

Hence, by putting $(\xi_n,\eta_n)=(1,0)$ into the formula (4.7), we obtain the result.

Now, we consider a natural extension of the algorithm (X,U). Let R be the subset of $\overline{X}$ such that

$$R = I_R \times [0,1) \cup I_{2,2} \times [0,1) \cup I_{2,R} \times [0,1/2)$$

and define the map U on R by

$$U(x,y) = \begin{cases} \left(k - \frac{1}{x}, \frac{1}{k-y}\right), & \text{if } (x,y) \in I_k \times [0,1), \\ \left(2 - \frac{1}{x}, \frac{1}{2-y}\right), & \text{if } (x,y) \in I_{2,2} \times [0,1), \\ \frac{-(k-2)+(k-1)x}{-(k-3)+(k-2)x}, \frac{(k-2)-(k-3)y}{(k-2)-(k-2)y}, & \\ & \text{if } (x,y) \in I_{2,k} \times [0,1/2). \end{cases}$$

Then we can easily see that

$$\bar{U}(I_k \times [0,1)) = \bar{T}(I_k \times [0,1)) = [0,1) \times I_k,$$

$$\bar{U}(I_{2,2} \times [0,1)) = \bar{T}(I_{2,2} \times [0,1)) = [0,1/2) \times [1/2,1),$$

$$\bar{U}(I_{2,k} \times [0,1/2)) = \bar{T}^{k-2}(I_{2,k} \times [0,1/2)) = [1/2,2/3) \times [\tfrac{k-2}{k-1}, \tfrac{k-1}{k}),$$

and

$$\bar{T}^j(I_{2,k} \times [0,1/2)) \cap R = \phi, \text{ if } 1 \leq j < k-2.$$

Therefore, let T_R be the induced automorphism of $\bar{T}$ on R, then

$$\bar{T}_k(x,y) = \bar{U}(x,y).$$

Hence, by theorem 3.1, the invariant measure $\bar{\nu}$ with respect to $\bar{U}$ is given as an induced measure on R of $\bar{\mu}$:

$$d\bar{\nu} = \frac{1}{2\log 2} \frac{d \times dy}{(1-xy)2},$$

where $2\log 2$ is a normalizing constant.

Taking the marginal distribution, we get a finite invariant measure ν with respect to U:

$$d\nu = \begin{cases} \frac{1}{2\log 2} \frac{dx}{1-x}, & \text{if } x \in [0,2/3), \\ \frac{1}{2\log 2} \frac{dy}{2-x}, & \text{if } x \in [2/3,1). \end{cases}$$

<u>Theorem 4.1</u>. *Let* $\bar{\nu}$ *and* ν *be as above, then the dynamical system* $(k, \bar{U}, \bar{\nu})$ *is the natural extension of the dynamical system* $(k, \bar{U}, \bar{\nu})$ *and the dynamical system* $(k, \bar{U}, \bar{\nu})$ *is the induced*

automorphism of $(\bar{X}, \bar{T}, \bar{\mu})$ *on* R.

Fundamental lemma 4.1. *For each irrational* $x \in X$, *we have*

$$\bar{U}^n(x,0) = (U^n x, -\frac{s_n}{r_n}) .$$

Proof. Put

$$\begin{pmatrix} r_n' & s_n' \\ t_n' & u_n' \end{pmatrix} = \begin{pmatrix} a_n & -1 \\ 1 & 0 \end{pmatrix} \cdots \begin{pmatrix} a_1 & -1 \\ 1 & 0 \end{pmatrix},$$

then from the definition of $\bar{U}$,

$$\bar{U}^n(x,y) = (U\, x, \frac{t_n' + u_n' \cdot y_n}{r_n' + s_n' \cdot y}) .$$

On the other hand, we see inductively that

$$r_n' = r_n,\ u_n' = u_n,\ t_n' = -s_n \text{ and } s_n' = -t_n,\ n \geq 1 .$$

Therefore, we see

$$\bar{U}^n(x,0) = (U^n x, \frac{t_n'}{r_n'}) = (U^n x, -\frac{s_n}{r_n}) .$$

Corollary 4.2. *For each irrational* $x \in X$,

(1) *if a simple fraction* $\frac{p}{q}$, $q > 0$, *satisfies the inequality*

$$0 < q(qx - p) < 1,$$

then there exists an integer k *such that*

$$\frac{p}{q} = \frac{t_k}{r_k},$$

(2) *if* $\frac{t_n}{r_n}$ *is the* n-th *convergent of* x, *which is induced by the algorithm* U, *then*

$$r_n(r_n \cdot x - t_n) \leq 2 .$$

Proof. By Remark 1.2, if a simple fraciton satisfies the assumption of (1), then there exists an integer n such that $\frac{p}{q} = \frac{p_n}{q_n}$. And the $\frac{p_n}{q_n}$ also satisfies the inequality $0 < q_n(q_n x - p_n) < 1$. Therefore, by Lemma 3.2, we see

$$\overline{T}^n(x,0) \quad D_1 .$$

For $\lambda = 1$, since D_1 is a subset of R, there exists an integer k such that

$$\overline{U}^k(x,0) = \overline{T}^n(x,0) .$$

Hence we know

$$\frac{p_n}{q_n} = \frac{t_k}{r_k} .$$

Part (2) is proved as follows. By Proposition 4.1, we see

$$r_n(I_n \cdot x - t_n) = \frac{U^n x}{(1 + \frac{s_n}{r_n})} = f(\overline{U}^n(x,0)) ,$$

where $f(x,y) = \frac{x}{1-xy}$. On the other hand, we know that

$$U(x,0) \quad R \text{ and } R \subset D_2 = \{(x,y);\ 0 < f(x,y) < 2\} .$$

Therefore, we obtain

$$r_n(r_n \cdot x - t_n) < 2 .$$

5. SOME METRICAL RESULTS

In this Section, the asymptotic behaviour of $\frac{t_n}{r_n}$ is discussed.

Proposition 5.1. *For almost all* $x \in (0,1)$,

$$-\lim_{n\to\infty} \frac{1}{n} \log (r_n \cdot x - t\ ) = \frac{\pi^2}{12\log 2} .$$

Proof. By Proposition 4.1, we know

$$-\frac{1}{n}\log(r_n\cdot x - t_n) = -\frac{1}{n}\sum_{k=0}^{n-1}\log g(U^R x) - \frac{1}{n}\log U^n x . \qquad (5.1)$$

By using an analogous method in the proof of Proposition 3.2 in [2], which is a kind of Borel-Canteli lemma, the second term $-\frac{1}{n}\log U^n x$ converges to zero as n tends to infinity for almost all x. The first term converges to the following constant c by an application of the ergodic theorem on (X, U, ν):

$$c = \int_X \log g(x)\, d\nu$$

$$= \frac{1}{2\log 2}\left[\int_0^{2/3}\log x \cdot \frac{dx}{(1-x)} + \sum_{k\geq 3}^{\infty}\log\{-(k-3)+(k-2)\cdot x\} \cdot \frac{dx}{2-x}\right] . \qquad (5.1)$$

The value of the constant c is calculated as

$$c = \frac{1}{2\log 2}\int_0^1 \log x \cdot \frac{dx}{(1-x)} = \frac{\pi^2}{12\log 2} ,$$

by using the change of variation of second terms in (5.1):

$$x = \frac{(k-2) - (k-3)z}{(k-1) - (k-2)z} \quad (k \geq 3) .$$

Proposition 5.2. *For almost all* $x \in (0,1)$,

(1) $\lim_{n\to\infty}\frac{1}{n}\log r_n = \frac{\pi^2}{12\log 2}$,

(2) $-\lim_{n\to\infty}\frac{1}{n}\log(x - \frac{t_n}{r_n}) = \frac{\pi^2}{6\log 2}$.

Proof. By Proposition 4.1, we have

$$\log r_n = -\log(r_n x - t_n) + \log f(U^n(x,0)) .$$

By using an analogous method in the proof of Proposition 3.2 in [2], we see

$$\lim_{n\to\infty} \frac{1}{n} \log f(U^n(x,0)) = 0 ,$$

for almost all $x \in X$. Therefore, by Proposition 5.4, $\frac{1}{n} \log r_n$ converges to the same constant.

Part (2) is obtained from Proposition 4.2 (1) and Proposition 5.2 (1).

Proposition 5.3. *For* $0 \leqq \lambda \leqq 1$,

$$\lim_{N\to\infty} \frac{1}{N} \#\{n: r_n(r_n x - t_n) < \lambda,\ 1 \leqq n \leqq N\} = \frac{\lambda}{2\log 2} ,$$

for almost all $x \in (0,1)$.

Proof. By the same manner as in Lemma 3.1, we know the equality:

$$\#\{n: r_n(r_n \cdot x - t_n) < \lambda,\ 1 \leqq n \leqq N\} = \sum_{n=1}^{N} I_\lambda(U^n(x,0)) ,$$

where I_λ is the indicator function of the set D_λ. By using the ergodic theorem, we see that for almost all $x \in X$

$$\frac{1}{N} \sum_{n=0}^{N-1} I_\lambda(\bar{U}^n(x,0)) \to \nu(D_\lambda) = \frac{\lambda}{2\log 2} .$$

(See the proof of theorem 3.1 in [2] in detail.)

Combining Propositions 5.1 and 5.3, we get the main result:

Theorem 5.1. *For* $0 \leqq \lambda \leqq 1$,

$$\lim_{N\to\infty} \frac{\#\{(q,p): 0 < q(qx - p) < \lambda,\ (q,p) = 1,\ 0 < q \leqq N\}}{\log N} = \frac{6}{\pi^2} \cdot \lambda ,$$

for almost all $x \in (0,1)$.

Proof. Choose an integer n satisfying $r_{n-1} \leqq N < r_n$, then by Corollary 4.2 (1) we have the equality:

$$\#\{(q,p): 0 < q(qx - p) < \lambda,\ (q,p) = 1,\ 0 < q \leqq N\}$$
$$= \#\{k: r_k(r_k \cdot x - t_k) < \lambda,\ k \leqq n-1\} .$$

Hence, by Propositions 5.3 and 5.1, we know

$$\lim_{N\to\infty} \frac{\#\{(q,p): 0 < q(qx - p) < \lambda,\ (q,p) = 1,\ 0 < q \leq N\}}{\log N}$$

$$\geq \lim_{n\to\infty} \frac{\#\{k: r_k(r_k \cdot x - t_k) < \lambda,\ k \leq n-1\} / n}{\log r_n / n}$$

$$= \frac{\dfrac{\lambda}{2\log 2}}{\dfrac{\pi^2}{12\log 2}} = \frac{6}{\pi^2} \cdot \lambda \text{ , for almost all } x \in (0,1) .$$

Replacing r_n by r_{n-1}, we obtain the reverse inequality.

REFERENCES

[1] Erdös, P.: Some results on Diophantine approximation, Acta Arith., 5, 359-369 (1959).

[2] Ito, Sh.: Algorithms with mediant convergents and their metrical theory, preprint.

[3] Ito, Sh. and Nakada, H.: On natural extensions of transformations related to Diophantine approximations, Number theory and Combinatrics, World Scientific Pub., 185-207 (1985).

[4] Takagi, T., Lectures on Elementary Number Theory, Second Ed., Kyoritsu Pub., Tokyo, 1971.

[5] Zagier, D.B., Zeta Funktionen und Quadratishe Körper, Springer, Berlin, 1981.

Proc. Prospects
of Math. Sci.
World Sci. Pub.
63-72, 1988

ON AN EXPONENTIAL SUM I

SHIGERU KANEMITSU

Department of Mathematics
Faculty of Science
Kyushu University
Fukuoka 812
JAPAN

ABSTRACT

For $N \in \mathbb{N}$ and $x > 0$ we consider the exponential sum

$$E_N(x) = \sum_{n \leq N} \exp(2\pi i x \sqrt{n})$$

and prove an asymptotic formula of the form

$$E_N(x) = \int_1^N \exp(2\pi i x \sqrt{t}) dt + \frac{1}{\sqrt{2}} x A(x) + O(\sqrt{x}+1)$$

for which $A(x) \neq o(1)$, debunking an optimistic view that $E_N(x)$ can be well approximated by the corresponding integral.

1. INTRODUCTION

In [2] an asymptotic formula for the sum $\sum_{n \leq N} B_r(\{\sqrt{n}\})$, where $B_r(x)$ is the r-th Bernoulli polynomial and $\{x\}$ is the fractional part of x, has been obtained on the basis of the formula

$$E_N(x) := \sum_{n \leq N} \exp(2\pi i x \sqrt{n}) = \int_1^N \exp(2\pi i x \sqrt{t})\, dt + O(x), \tag{1}$$

valid uniformly in N. This can be proved by the Euler-Maclaurin sum formula as noted by S. Uchiyama (cf. [6], Section 1). The first object of the present paper is to widen the range of validity of Kano's theorem, which has been established in [3] (unpublished). The second is a refinement of (1) in the case where $x \leq \sqrt{c_1 N}$ holds with some constant c_1, $0 < c_1 < 4$, namely, we shall prove in Section 2 the following

Theorem. *Let* N *be a positive integer and* x *be a positive real, and define*

$$E_N(x) = \sum_{n \leqq N} \exp(2\pi i x \sqrt{n}).$$

Suppose that the condition

$$\text{(I)} \qquad 0 < x < \sqrt{c_1 N}$$

holds with some absolute constant c_1 *satisfying* $0 < c_1 < 4$. *Then*

$$E_N(x) = \int_1^N \exp(2\pi i x \sqrt{t})dt + \frac{1}{\sqrt{2}} x A(x) + O(\sqrt{x+1}) , \qquad (2)$$

where

$$A(x) = \sum_{m=1}^{\infty} m^{-3/2} \exp\left\{ i(\frac{\pi x^2}{2m} - \frac{\pi}{4}) \right\} ; \qquad (3)$$

and if either of the following two conditions is satisfied:

$$\text{(II,1)} \qquad \begin{array}{l} \sqrt{c_1 N} < x \leqq 2^{1/(2-\lambda)} N^{3/(4-2\lambda)} , \\ \left| \left\{ \frac{x}{2\sqrt{N}} \right\} - \frac{1}{2} \right| \leqq c_2 x^{2-\lambda} N^{-3/2} \end{array}$$

for some constants λ, c_2 *such that* $0 \leqq \lambda < 1$, $0 < c_1 < 1$, *or*

$$\text{(II,2)} \qquad 2^{1/(2-\lambda)} N^{3/(4-2\lambda)} < x ,$$

then

$$E_N(x) = O(x^{1/2} N^{1/4}) + O(x^{1-\lambda}) . \qquad (4)$$

2. PROOF OF THE THEOREM

To prove the theorem we need the following lemmas.

Lemma 1 ([5], Lemma 4.3). *If* f(x) *and* g(x) *are real functions such that* $f'(x)/g(x)$ *is monotonic, and* $f'(x)/g(x) \geqq m > 0$, *or* $\leqq -m < 0$ *over* [a,b], *then*

$$\left| \int_a^b g(t) \exp(if(t))\, dt \right| \le \frac{4}{m} .$$

Lemma 2 (cf. [3], Lemma 1). *If* $\alpha > 0$ *is a constant, then*

(i) $$\sum_{n=1}^{\infty} \frac{1}{n(n+\alpha)} = \frac{1}{\alpha} \log(\alpha+1) + O(\frac{1}{\alpha+1}) ;$$

(ii) $$\sum_{n>\alpha} \frac{1}{n(n-\alpha)} = \frac{1}{\alpha} \log(\alpha+1) + O(\frac{1}{\alpha(1-\{\alpha\})})$$

for $1 < \alpha \notin \mathbf{Z}$;

(iii) $$\sum_{n<\alpha} \frac{1}{n(\alpha-n)} = \frac{2}{\alpha} \log \alpha + O(\frac{1}{\alpha\{\alpha\}})$$

for $1 < \alpha \notin \mathbf{Z}$.

Proof. (i) and (ii) follow immediately from the following proposition by taking $f(t) = \frac{1}{t(t+\alpha)}$, $f(t) = \frac{1}{(t+[\alpha])(t-\{\alpha\})}$: If $f(x) \geqq 0$ is defined for $x \geqq 1$ such that $f(x)$ is monotonously decreasing and $\int_1^{\infty} f(t)\, dt$ exists, then $\sum_{n=1}^{\infty} f(n)$ exists, and

$$\int_1^{\infty} f(t)\, dt \le \sum_{n=1}^{\infty} f(n) \le f(1) + \int_1^{\infty} f(t)\, dt .$$

As regards (iii), we proceed as follows: We have

$$\sum_{n<\alpha} \frac{1}{n(\alpha-n)} = \frac{1}{\alpha} \sum_{n=1}^{[\alpha]} \frac{1}{n} + \frac{1}{\alpha} \sum_{n=1}^{[\alpha]} \frac{1}{[\alpha]+1-n}$$
$$+ \frac{1}{\alpha} \sum_{n=1}^{[\alpha]} \frac{1-[\alpha]}{(\alpha-n)([\alpha]+1-n)} ,$$

and note that the last summand belongs to $(0, \frac{1}{\{\alpha\}})$, and that $\log\,[\alpha] = \log \alpha + O(1)$. Thus (iii) follows, thereby proving our Lemma.

Proof of the theorem. If (II,1) or (II,2) holds, then we put

$$\eta = \eta(x,N) = \frac{4N^{3/2}}{x}(\left\{\frac{x}{2\sqrt{N}}\right\} - \frac{1}{2}) . \tag{5}$$

If (II,2) holds, then $\left| \left\{ \frac{x}{2\sqrt{N}} \right\} - \frac{1}{2} \right| \leq \frac{1}{2} < \frac{x^{2-\lambda}}{4N^{3/2}}$,

so that *If is a constant then*

$$\eta < \min\left(\frac{2N^{3/2}}{x}, x^{1-\lambda}\right) ; \tag{6}$$

If (II,1) holds, then

$$\eta < \min\left(\frac{2N^{3/2}}{x}, 4c_2 x^{1-\lambda}\right) . \tag{7}$$

for

Furthermore, we define

for

$$\delta = \delta(x,N) = \begin{cases} \frac{N}{x^2} + \frac{\eta}{x^2} & \text{if (II,1) or (II,2) holds,} \\ c_1^{-1} & \text{if (I) holds,} \end{cases} \tag{8}$$

$$\varepsilon = \varepsilon(x) = \frac{1}{(2[x] + 3)^2} . \tag{9}$$

Then by the binomial expansion ($|\theta| \leq 1$)

$$\frac{1}{2\sqrt{\delta}} = \frac{x}{2\sqrt{N+\eta}} = \left[\frac{x}{2\sqrt{N}}\right] + \frac{1}{2} + \left\{\frac{x}{2\sqrt{N}}\right\} - \frac{1}{2} - \frac{x\eta}{4N^{3/2}}$$

$$+ \frac{3}{8}\,\theta\,\frac{x}{N^{5/2}}\,\eta^2 .$$

If (II,2) holds, then by (6)

$$\frac{1}{2\sqrt{\delta}} = \frac{1}{2} + \frac{3}{2}\,\theta\,2^{-1/3}\,x^{-(1+\lambda)/3} = \frac{1}{2} + o(1)$$

for N (therefore for x) large enough.

If (II,1) holds, then

$$\left\{\frac{1}{2\sqrt{\delta}}\right\} = \frac{1}{2} + 6c_2^2\theta\, x^{3-2\lambda}N^{-5/2} = \frac{1}{2} + \frac{\theta'}{4} .$$

Thus, in any case

$$\left\{\frac{1}{2\sqrt{\delta}}\right\},\ 1-\left\{\frac{1}{2\sqrt{\delta}}\right\} \gg 1\,,$$
$$\left\{\frac{1}{2\sqrt{\varepsilon}}\right\},\ 1-\left\{\frac{1}{2\sqrt{\varepsilon}}\right\} \gg 1\,. \tag{10}$$

It is clear from (9) that

$$\sum_{n\leqq\varepsilon x^2} \exp(2\pi i x\sqrt{n}) = 0\,. \tag{11}$$

Also if (I) holds, then van der Corput's lemma applied to the sum $\sum_{\frac{x^2}{c_1}<n\leqq N} \exp(2\pi i x\sqrt{n})$ will give rise to

$$\sum_{\frac{x^2}{c_1}<n<N} \exp(2\pi i x\sqrt{n}) = \int_{\frac{x^2}{c_1}}^{N} \exp(2\pi i x\sqrt{t})\,dt + O(1)\,. \tag{12}$$

By (11), (12) and $\sum_{n=N}^{x^2} = O(x^{1-\lambda})$ in Case II, it suffices to treat the sum

$$E_N'(x) = \sum_{\varepsilon x^2<n\leqq\delta x^2} \exp(2\pi i x\sqrt{n})\,.$$

We shall deal with the real part $S_N'(x)$ of $E_N'(x)$ only, since the argument will proceed on the same lines as regards the imaginary part.

By the Euler-Maclaurin sum formula we have

$$\begin{aligned} S_N'(x) &= \sum_{\varepsilon x^2<n\leqq\delta x^2} \cos(2\pi x\sqrt{n}) \\ &= \int_{\varepsilon x^2}^{\delta x^2} \cos(2\pi x\sqrt{u})\,du + xJ(x) + O(1)\,, \end{aligned} \tag{13}$$

where

$$J(x) = -\pi\int_{\varepsilon x^2}^{\delta x^2} P_1(u)\,\frac{\sin(2\pi x\sqrt{u})}{\sqrt{u}}\,du\,, \tag{14}$$

$P_1(u) = u - [u] - \frac{1}{2}$ being the saw-tooth function.

Using the Fourier series for $P_1(u)$ and making the substitution

$u = t^2$, we have

$$J(x) = 2\int_{\sqrt{\varepsilon x}}^{\sqrt{\delta x}} \sum_{m=1}^{\infty} \frac{\sin(2\pi m t^2)}{m} \sin(2\pi x t)\, dt$$

$$= \sum_{m=1}^{\infty} \frac{1}{m} \int_{\sqrt{\varepsilon x}}^{\sqrt{\delta x}} 2\sin(2\pi m t^2)\sin(2\pi x t)\, dt$$

$$= \sum_{m=1}^{\infty} \frac{1}{m} I^- - \sum_{m=1}^{\infty} \frac{1}{m} I^+ , \tag{15}$$

termwise differentiation here being justified by Lebesgue's dominated convergence theorem, and I^+, I^- denoting

$$I^{\pm} = I^{\pm}(\varepsilon,\delta,m,x) = \int_{\sqrt{\varepsilon x}}^{\sqrt{\delta x}} \cos 2\pi(mt^2 \pm xt)\, dt , \tag{16}$$

where the signs $\pm$ should be taken in this order (also in the sequel).

We shall now consider the integrals $I^{\pm}$. Making the substitution $t = \frac{1}{m}(v \mp \frac{x}{2})$ in $I^{\pm}$ accordingly, we have

$$I^{\pm} = \frac{1}{m} \int_{(m\sqrt{\varepsilon}\pm 1/2)x}^{(m\sqrt{\delta}\pm 1/2)x} \cos \frac{2\pi}{m} (v^2 - \frac{x^2}{4})\, dv .$$

We estimate I^+ by means of Lemma 1 to get

$$I^+ = O(x^{-1}(\sqrt{\varepsilon}m + \tfrac{1}{2})^{-1}) .$$

Consequently,

$$\sum_{m=1}^{\infty} \frac{1}{m} I^+ = O(\frac{1}{x\sqrt{\varepsilon}} \sum_{m=1}^{\infty} \frac{1}{m(m + \frac{1}{2\sqrt{\varepsilon}})} = O(\frac{\log x}{x}) \tag{17}$$

by Lemma 2, (i).

As regards $\sum_{m=1}^{\infty} \frac{1}{m} I^-$ we truncate it by applying Lemma 1 to those summands with $m > \frac{1}{2\sqrt{\varepsilon}}$. Indeed, by Lemma 2, (ii) and (10)

$$\sum_{m>\frac{1}{2\sqrt{\varepsilon}}} \frac{1}{m} I^- = O\left(\frac{1}{x\sqrt{\varepsilon}} \sum_{m>\frac{1}{2\sqrt{\varepsilon}}} \frac{1}{m(m - \frac{1}{2\sqrt{\varepsilon}})}\right)$$

$$= O\left(\frac{\log x}{x}\right) .$$

Hence

$$\sum_{m=1}^{\infty} \frac{1}{m} I^- = \sum\nolimits_1 + \sum\nolimits_2 + O\left(\frac{\log x}{x}\right) , \tag{18}$$

where

$$\sum\nolimits_1 = \sum_{m<\frac{1}{2\sqrt{\delta}}} \frac{1}{m} I^- , \quad \sum\nolimits_2 = \sum_{\frac{1}{2\sqrt{\delta}}<m<\frac{1}{2\sqrt{\varepsilon}}} \frac{1}{m} I^-$$

(if (I) holds, then $\sum_1 = 0$).

We consider I^- for this range of m. We write

$$I^- = \frac{1}{m} \cos\left(\frac{\pi x^2}{2m}\right)C(m,x) + \sin\left(\frac{\pi x^2}{2m}\right)S(m,x) , \tag{19}$$

where

$$\begin{matrix} C \\ S \end{matrix}(m,x) = \begin{matrix} C \\ S \end{matrix}(\varepsilon,\delta,m,x) = \int_{(\sqrt{\varepsilon}m-1/2)x}^{(\sqrt{\delta}m-1/2)x} \begin{matrix} \cos \\ \sin \end{matrix} \left(\frac{2\pi}{m} v^2\right) dv .$$

Let us treat $\sum_2$ first. Making the substitution $v = \sqrt{m}\, u$ in $\begin{matrix} C \\ S \end{matrix}(m,x)$, we may write

$$\begin{matrix} C \\ S \end{matrix}(m,x) = \sqrt{m} \left\{ \int_0^{(1/2-\sqrt{\varepsilon}m)x/\sqrt{m}} \begin{matrix} \cos \\ \sin \end{matrix}(2\pi u^2)\, du \right.$$

$$\left. + \int_0^{(\sqrt{\delta}m-1/2)x/\sqrt{m}} \begin{matrix} \cos \\ \sin \end{matrix}(2\pi u^2)\, du \right\} .$$

Since for $T > 0$ we have

$$\int_0^T \begin{matrix} \cos \\ \sin \end{matrix}(2\pi u^2)\, du = \frac{1}{4} + O\left(\frac{1}{T}\right) ,$$

we find

$$\begin{matrix} C \\ S \end{matrix}(m,x) = \sqrt{m} \left\{ \frac{1}{2} + O\left(\frac{\sqrt{m}}{(\frac{1}{2} - \sqrt{\varepsilon}m)x}\right) + O\left(\frac{\sqrt{m}}{(\sqrt{\delta}m - \frac{1}{2})x}\right) \right\} ,$$

whence follows by dint of (19) that

$$\sum_2 = \frac{1}{\sqrt{2}} \sum_{\frac{1}{2\sqrt{\delta}}<m<\frac{1}{2\sqrt{\varepsilon}}} m^{-3/2} \cos\left(\frac{\pi x^2}{2m} - \frac{\pi}{4}\right)$$

$$+ O\left(\frac{1}{x\sqrt{\varepsilon}} \sum_{\frac{1}{2\sqrt{\delta}}<m<\frac{1}{2\sqrt{\varepsilon}}} \frac{1}{m\left(\frac{1}{2\sqrt{\varepsilon}} - m\right)}\right)$$

$$+ O\left(\frac{1}{x\sqrt{\delta}} \sum_{\frac{1}{2\sqrt{\delta}}<m<\frac{1}{2\sqrt{\varepsilon}}} \frac{1}{m\left(m - \frac{1}{2\sqrt{\delta}}\right)}\right) .$$

Since

$$\sum_{\frac{1}{2\sqrt{\delta}}<m<\frac{1}{2\sqrt{\varepsilon}}} \frac{1}{m\left(\frac{1}{2\sqrt{\varepsilon}} - m\right)} \leq \sum_{m<\frac{1}{2\sqrt{\varepsilon}}} \frac{1}{m\left(\frac{1}{2\sqrt{\varepsilon}} - m\right)}$$

$$= O\left(\sqrt{\varepsilon}\log\left(-\frac{1}{2\sqrt{\varepsilon}}\right)\right) + O\left(\left\{\frac{1}{2\sqrt{\varepsilon}}\right\}^{-1}\right) ,$$

and

$$\sum_{\frac{1}{2\sqrt{\delta}}<m<\frac{1}{2\sqrt{\varepsilon}}} \frac{1}{m\left(m - \frac{1}{2\sqrt{\delta}}\right)} \leq \sum_{\frac{1}{2\sqrt{\delta}}<m} \frac{1}{m\left(m - \frac{1}{2\sqrt{\delta}}\right)}$$

$$= O\left(\sqrt{\delta}\log\left(\frac{1}{2\sqrt{\delta}} + 1\right)\right) + O\left(\sqrt{\delta}\,\frac{1}{1 - \left\{\frac{1}{2\sqrt{\delta}}\right\}}\right)$$

by Lemma 2, (iii) and (ii), respectively, we conclude that

$$\sum_2 = \frac{1}{\sqrt{2}} \sum_{\frac{1}{2\sqrt{\delta}}<m<\frac{1}{2\sqrt{\varepsilon}}} m^{-3/2} \cos\left(\frac{\pi x^2}{2m} - \frac{\pi}{4}\right) + O\left(\frac{\log x}{x}\right) . \qquad (20)$$

Lastly we estimate $\sum_1$ (only if case (II) holds). Since $\sqrt{\delta}m - \frac{1}{2} < 0$, we have by Lemma 1 and Lemma 2, (iii)

$$\sum_1 = \sum_{m<\frac{1}{2\sqrt{\delta}}} m^{-2} \int_{(1/2-\sqrt{\delta}m)x}^{(1/2-\sqrt{\varepsilon}m)x} \cos \frac{2\pi}{m}\left(v^2 - \frac{x^2}{4}\right) dv$$

$$= O\left(\frac{\log x}{x}\right) \qquad (21)$$

in view of (8).

It follows from (18), (20), (21) that

$$\sum_{m=1}^{\infty} \frac{1}{m} I^{-} = \frac{1}{\sqrt{2}} \sum_{\frac{1}{2\sqrt{\delta}}<m<\frac{1}{2\sqrt{\varepsilon}}} m^{-3/2} \cos\left(\frac{\pi x^2}{2m} - \frac{\pi}{4}\right) + O\left(\frac{\log x}{x}\right)$$

$$= \begin{cases} \frac{1}{\sqrt{2}} \operatorname{Re}A(x) + O\left(\frac{\log x}{x}\right) + O(\varepsilon^{-1/4}) , & \text{(I)} \\ O(\delta^{-1/4}) + O\left(\frac{\log x}{x}\right) , & \text{(II)} \end{cases}$$

$$= \begin{cases} \frac{1}{\sqrt{2}} \operatorname{Re}A(x) + O\left(\frac{\log x}{x}\right) + O\left(\frac{1}{\sqrt{x}}\right) , & \text{(I)} \\ O\left(\frac{N^{1/4}}{\sqrt{x}}\right) . & \text{(II)} \end{cases}$$

Hence

$$E_N(x) = \int_1^N \exp(2\pi i x\sqrt{u})du + \begin{cases} \frac{1}{\sqrt{2}} x\, A(x) + O(\sqrt{x}) , & \text{(I)} \\ O(x^{1-\lambda}) + O(\sqrt{x}N^{1/4}) , & \text{(II)} \end{cases}$$

$$= \begin{cases} \int_1^N \exp(2\pi i x\sqrt{u})\, du + \frac{1}{\sqrt{2}} x\, A(x) + O(\sqrt{x}) , & \text{(I)} \\ O(x^{1-\lambda}) + O(\sqrt{x}N^{1/4}) . & \text{(II)} \end{cases}$$

This completes the proof.

Remark. As regards the function $A(x)$ given by (3) we note that

$$\limsup_{x \to \infty} |A(x)| > \frac{1}{\sqrt{2}} \zeta\left(\frac{3}{2}\right) \quad (> 1) \tag{22}$$

holds. In fact, for each $k \in \mathbb{N}$, we have

$$|A(2\sqrt{k!})| \geq \left| \sum_{n \leqq k} n^{-3/2} \exp\left(\frac{2k!\pi}{n} - \frac{\pi}{4}\right) \right| - \sum_{n>k} n^{-3/2}$$

$$= \frac{1}{\sqrt{2}} \sum_{n \leqq k} n^{-3/2} + O\left(\frac{1}{\sqrt{k}}\right)$$

$$\to \frac{1}{\sqrt{2}} \zeta\left(\frac{3}{2}\right) , \qquad \text{as } k \to \infty ,$$

which proves (22).

Concluding Remarks. Our theorem is a negative theorem in its nature in the sense that it says that one cannot replace $O(x)$ in (1) by $o(x)$. To save this defect there are some ways: to find a more appropriate main term; to take an average of $E_N(x)$ as in Uchiyama [6], [7] or to consider a metrical version similar to Erdös' results [1]. In the forthcoming part II we shall consider these problems, taking into account Potockii's result [4].

REFERENCES

[1] Erdös, P.: Some remarks on Diophantine approximations, J. Indian Math. Soc., (N.S.) 12, 67-74 (1948).

[2] Kanemitsu, S. and R. Sita Rama Chandra Rao: On a paper of S. Chowla and H. Walum, Norske Vid. Selsk. Skr., 1, 1-3 (1982).

[3] Kano, T., Some Applications of Exponential Sums, Doctoral thesis, Rikkyo Univ., Japan, 1976.

[4] Potockii, V. V.: A sharpening of the estimate of a certain trigonometrical sum, Izv. Vyss. Uceb. Zaved. Matematika No.3, 82, 42-51 (1969).

[5] Titchmarsh, E. C., The Theory of the Riemann Zeta-function, Oxford University Press, Oxford, 1951.

[6] Uchiyama, S.: On some exponential sums involving the divisor functions, J. Reine Angew. Math., 262/263, 248-260 (1973).

[7] Uchiyama, S.: On some exponential sums involving the divisor function over arithmetical progressions, Math. J. Okayama Univ., 137-146 (1973/74).

Proc. Prospects
of Math. Sci.
World Sci. Pub.
73-77, 1988

A NOTE ON THE FIRST CASE OF FERMAT'S LAST THEOREM

YOSHIKAZU KARAMATSU

Department of Mathematics
Utsunomiya University
350 Mine-Machi, Utsunomiya
Japan

1. INTRODUCTION

Let p be an odd prime and B_n be the n-th Bernoulli number defined by the formal power series expansion $x/(e^x-1) = \sum_{k=0}^{\infty}(B_k/k!)x^k$.

It is well-known that if the equation

$$x^p + y^p + z^p = 0, \quad p \nmid xyz, \tag{1}$$

is satisfied in non-zero integers x, y and z prime to each other, then

$$\begin{aligned} &f_{p-1}(t) \equiv 0 \pmod{p}, \\ &B_{2m}f_{p-2m}(t) \equiv 0 \pmod{p}, \quad m = 1, 2, \cdots, (p-3)/2, \end{aligned} \tag{2}$$

where $-t \equiv x/y,\ y/x,\ x/z,\ z/x,\ y/z,\ z/y \pmod{p}$ and

$$f_n(X) = \sum_{k=1}^{p-1} k^{n-1}X^k .$$

The congruences (2) are so-called the Kummer-Mirimanoff Criteria. In this note we shall survey the results related to the stably irregularity index of p, under the assumption for which the equation (1) holds.

2. INTENSIVE RESULTS

In 1905 Mirimanoff[12] showed by using (2) that if (1) holds, then

$$B_{p-1-2n} \equiv 0 \pmod{p} , \tag{3}$$

for $n = 1, 2, 3, 4$. After that, this result has been extended by various authors. See Morishima[13] and Lehmer[10] for $n = 5, 6$; Karamatsu[5, 6, 7] and Wada[18] for $n = 7, 8, 9$; Keller and Löh[8], and Müller[14] for $n = 10, 11, \cdots, 22$.

On the other hand, Krasner[9] has given a remarkable result in 1934:

Theorem 1. *If* (1) *is satisfied with the exponent* $p > (45!)^{88}$, *then* (3) *holds for* $n = 1, 2, \cdots, [\sqrt[3]{\log p}]$, *where* $[\alpha]$ *is the greatest integer in* α .

If the congruence (3) holds for $n = 1, 2, \cdots, \delta(p)$, however $B_{p-1-2(\delta(p)+1)} \not\equiv 0 \pmod{p}$, then the number $\delta(p)$ is called the stably irregularity index of p .

Since $22 < \sqrt[3]{\log(45!)^{88}} < 23$ (cf. Keller and Löh[8]), so it may be deduced that

Theorem 2 (Karamatsu[7]), Keller and Löh[8]). *If* (1) *is satisfied*, *then* $\delta(p) \geqq \max\{22, [\sqrt[3]{\log p}]\}$.

Recently, Sami[17] and Granville[3] have improved this result. We would like to present Granvill's result here:

Theorem 3 (Granville[3]). *If* (1) *is satisfied*, *then*

$$\delta(p) \geqq [(\log p / \log\log p)^{1/2}] .$$

If (3) holds for an integer n with $(p-3)/2 \geqq n \geqq 1$, then (Ribenboim[15,16])

$$\sum_{j=1}^{\alpha(p)} \frac{1}{j^{2n+1}} \equiv 0 \pmod{p} , \tag{4}$$

for $\alpha(p) = [p/3], [p/6]$.

Hence, using this fact we can state from Theorem 3 that:

Theorem 4. *If* (1) *is satisfied, then* (4) *holds for* $n = 1, 2, \cdots, [(\log p / \log\log p)^{1/2}]$.

Proof. The equation (1) has no solutions for $p = 3$ and 5. So assume that $p \geqq 7$. We know the following E. Lehmer's congruences [4, 11]: If $p-1 \nmid 2k-2$, then

$$(3^{2k}-1)\frac{B_{2k}}{4k} \equiv \sum_{0<a<p/3} (p-6a)^{2k-1} \pmod{p^2}, \tag{5}$$

$$(6^{2k-1}+ 3^{2k-1}+ 2^{2k-1}- 1)\frac{B_{2k}}{4k} \equiv \sum_{0<a<p/6} (p-6a)^{2k-1} \tag{6}$$

$$\pmod{p^2},\ p \geqq 7 .$$

We can deduce from the congruences (5) and (6), respectively

$$(3^{2k}-1)\frac{B_{2k}}{4k} \equiv (-3)^{2k-1} \sum_{0<a<p/3} a^{2k-1} \pmod{p}, \tag{5a}$$

and

$$(6^{2k-1}+ 3^{2k-1}+ 2^{2k-1}- 1)\frac{B_{2k}}{4k}$$

$$\equiv (-6)^{2k-1} \sum_{0<a<p/6} a^{2k-1} \pmod{p} . \tag{6a}$$

Now, assuming $B_{p-1-2n} \equiv 0 \pmod{p}$ and taking $2k = p-1-2n$, we then obtain from (5a) and (6a) that

$$\sum_{a=1}^{\alpha(p)} a^{p-2-2n} \equiv \sum_{a=1}^{\alpha(p)} \frac{1}{a^{2n+1}} \equiv 0 \pmod{p},$$

for $\alpha(p) = [p/3], [p/6]$.

Hence, from Theorem 3, we obtain (4) for $n = 1, 2, \cdots, [(\log p / \log\log p)^{1/2}]$.

Let $\gamma(p)$ be the irregularity index of p, i.e.,

$$\gamma(p) = \#\{\ 2k \mid B_{2k} \equiv 0 \pmod{p},\ 2 \leqq 2k \leqq p-3\ \} .$$

Brückner's result[2] asserts that if (1) is satisfied, then $\gamma(p) \geqq [\sqrt{p}] - 2$. Therefore, Theorem 3 is weaker than Brückner's, insofar as we discuss (1) from the viewpoint of the irregularity index. However, we emphasize here that the conclusion of Theorem 3 is an unlikely event, since (3) must be satisfied for at least $[(\log p / \log\log p)^{1/2}]$ consecutive integers n.

Also, recently Agoh[1] showed the following interesting theorem:

Theorem 5 (Agoh). *Let* $p \geqq 5$ *be a prime and*

$$N = [\{\log(p+1) - \log 10\} / \log 3] + 2 .$$

If (1) *holds, then there exist at least* N *even integers* $k_1, k_2, \cdots, k_N$ *such that* $2[(p-1)/4] \geqq k_1 \geqq 2$, $2[(k_{i-1} + 2p)/6] \geqq k_i \geqq k_{i-1} + 2$ $(i = 2, 3, \cdots, N)$ *and* $B_{k_i} \equiv 0 \pmod p$ *for all* $i = 2, 3, \cdots, N$.

These last inequalities for indices of Bernoulli numbers seem to be of vital importance for the study of irregular pairs in certain intervals.

REFERENCES

[1] Agoh, T.: On the first case of Fermat's last theorem II, Manuscripta Math., 56, 465-474 (1986).

[2] Brückner, H: Zum ersten Fall der Fermatschen Vermutung, J. reine angew. Math., 274/275, 21-26 (1975).

[3] Granville, A.: On Krasner's criteria for the first case of Fermat's last theorem, Manuscripta Math., 56, 67-70 (1986).

[4] Johnson, W.: p-adic proofs of congruences for the Bernoulli numbers, J. of Number Theory, 7, 251-265 (1975).

[5] Karamatsu, Y. and Abe, S.: On Fermat's last theorem and the first factor of the class number of the cyclotomic field, TRU Math., 4, 1-9 (1968).

[6] Karamatsu, Y.: On Fermat's last theorem and the first factor of the class number of the cyclotomic field II, TRU Math., 16, 23-29 (1980).

[7] Karamatsu, Y.: Ribenboim's criteria and same criteria for the first case of Fermat's last theorem, TRU Math., 17, 25-38 (1981).

[8] Keller, W. and Löh, G.: The Criteria of Kummer and Mirimanoff extended to include 22 consecutive irregular pairs, Tokyo J. Math., 6, 397-402 (1985) and Supplement p.487.

[9] Krasner, M.: Sur le premier cas du théorème de Fermat, C. R. Acad. Sci. Paris, 199, 256-258 (1934).

[10] Lehmer, D. H.: A note on Fermat's last theorem, Bull. Amer. Math. Soc., 38, 723-724 (1932).

[11] Lehmer, E.: On congruences involving Bernoulli numbers and the quotients of Fermat and Wilson, Ann. Math., 39, 350-360 (1938).

[12] Mirimanoff, D.: L'équation indéterminée $x^p + y^p + z^p = 0$ et le critérium de Kummer, J.reine angew. Math., 128, 45-68 (1905).

[13] Morishima, T.: Über die Fermatsche Vermutung VII, Proc. Imperial Acad. Japan, 8, 63-66 (1932).

[14] Müller, H.: On some congruences concerning the criteria of Kummer, Expositiones Math., 2, 85-89 (1984).

[15] Ribenboim, P.: Some criteria for the first case of Fermat's last theorem, Tokyo J. Math., 1, 149-155 (1978).

[16] Ribenboim, P., 13 Lectures on Fermat's Last Theorem, Springer, New York-Heidelberg-Berlin, 1979.

[17] Sami, Z.: On the first case of Fermat's last theorem, Glasnik Mathematicki, 21, 259-269 (1986).

[18] Wada, H.: Some computations on the criteria of Kummer, Tokyo J. Math., 3, 173-176 (1980).

Proc. Prospects
of Math. Sci.
World Sci. Pub.
79-96, 1988

MONOPOLES WITH ONE-POINT SINGULARITY*

TOSHIYUKI MAEBASHI
Department of Mathematics
Faculty of Sciences
Kumamoto University
Kurokami, Kumamoto 860
Japan

1. INTRODUCTION

The monopole on Eucliden n-space E^n is a solution of the Yang-Mills equations:

$$DF = 0, \quad D{*}F = 0 \tag{1.1}$$

with the conditions:

$$\int_{E^n} F\ {*}F < +\infty\ , \tag{1.2}$$

$$\text{smoothness on } E^n\ . \tag{1.3}$$

It is a well-known fact that there exist no such solutions for n>4 [1]. We, however, can assert that we can find several solutions if we drop the above conditions. For the Laplace equation there exist interesting solutions with one point singularity such as $\frac{xdy-ydx}{r^2}$ on E^2 $((x,y)\in E^2,\ r^2=x^2+y^2)$, $\frac{1}{r}$ on E^3. The objective of this paper is to present such an example for the Yang-Mills equations on E^n. The solution is close to of finite action in the sense that, a small ball with the singular point as center being removed, the action turns out finite. We can also show the existence of a solution with two

---

* This paper is based on part of my talk entitled "Dirac monopoles and t'Hooft monopoles" at the symposium "Prospects of Mathematical Science."

point singularity for the Yang-Mills equations on n-sphere $S^n \subset E^{n+1}$, which is related to the above solution on E^n through the stereographic projection.

Let us go on into some details. Let x be a point of E^n and $x_1, \cdots, x_n$ the coordinate components of x. We write

$$r = \sqrt{x_1^2 + \cdots + x_n^2}\ .$$

Then our solution with a singularity at the origin is given by an $\mathfrak{so}(n)$-valued differential form:

$$\omega = \frac{1}{r^2}\begin{pmatrix} 0 & x_1dx_2 - x_2dx_1 & \cdots & x_1dx_n - x_ndx_1 \\ x_2dx_1 - x_1dx_2 & 0 & \cdots & x_2dx_n - x_ndx_2 \\ \cdot & \cdot & \cdot & \cdot \\ x_ndx_1 - x_1dx_n & & \cdots & 0 \end{pmatrix}.$$

We further find that the curvature from F turns out to be

$$\frac{1}{r^4}\, C\ ,$$

where C is the $n\times n$ matrix the (i,j)-component of which is given by

$$C_{ij} = \sum_{k\neq i,j} x_k(x_i dx_j \wedge dx_k + x_j dx_k \wedge dx_i + x_k dx_i \wedge dx_j).$$

Let us consider the case where $n=3$ and write $x=x_1$, $y=x_2$, $z=x_3$. We set

$$\Phi = \frac{1}{r}\begin{pmatrix} 0 & z & -y \\ -z & 0 & x \\ y & -x & 0 \end{pmatrix}.$$

This is a Higgs field satisfying $D\Phi = 0$ [2]. We have

$$F = \frac{1}{r^3}(x dy \wedge dz + y dz \wedge dx + z dx \wedge dy)\Phi\ ,\quad *F = \frac{x dx + y dy + z dz}{r^3}\,\Phi\ .$$

We discuss the geometric meanings of the connection in Appendix 2.

The author would like to acknowledge his sincere thanks to the referees. Especially the geometric observation in Appendix 2 is entirely due to one of the referees.

2. RUDIMENTS OF CONNECTIONS

2.1 Surfaces in E^3

Tangent bundle connection

Let X be an oriented surface in Euclidean 3-space E^3 and T_x be the tangent space at x of X. Take a frame $e=(e_1,e_2)$ for T_x with

$$(e_i,e_j) = \delta_{ij} , \quad i,j = 1,2 , \tag{2.1.1}$$

where the latter (,) stands for the inner product of E^3. Let ξ be the normal at x of X. We express de_i with respect to e_1, e_2, ξ:

$$de_i = \sum_{j=1}^{2} \omega_{ji} e_j + \Omega_i \xi ,$$

then the absolute differentials De_i are $\sum_{j=1}^{2} \omega_{ji} e_j$. From (2.1.1) we have $\omega_{12}=-\omega_{21}$. Hence the "connection form"

$$\omega = \begin{pmatrix} 0 & \omega_{12} \\ \omega_{21} & 0 \end{pmatrix}$$

may be considered as a differential 1-form with its values in the Lie algebra $\mathfrak{so}(2)$ of $SO(2)$. Let $\eta=\sum \eta_i e_i$ be any tangent vector field on X. Then the absolute differential of η is given by

$$D_\eta = e \left\{ \begin{pmatrix} d\eta_1 \\ d\eta_2 \end{pmatrix} + \omega \begin{pmatrix} \eta_1 \\ \eta_2 \end{pmatrix} \right\} .$$

Taking another frame $e'=(e'_1,e'_2)$ with $e'=eg$, $g \in SO(2)$, we have

$$e'\omega' = De' = Deg + edg = e(\omega g + dg),$$

whence

$$\omega' = g^{-1}\omega g + g^{-1}dg .$$

The curvature form F, which is defined as $F = d\omega + \omega \wedge \omega$, has the transformation law:

$$F' = g^{-1}Fg .$$

2.2 The Riemann Sphere

Tangent bundle connection

We can introduce a complex coordinate into 2-sphere S^2 (except for the north pole) by

$$w = \frac{x+\sqrt{-1}y}{1-z} , \tag{2.2.1}$$

where $(x,y,z) \in S^2$. On the other hand we have the polar coordinates Θ, φ on S^2 such that

$$w = \frac{\sin\varphi e^{\sqrt{-1}\Theta}}{1-\cos\varphi} .$$

Then

$$dw = \frac{e^{\sqrt{-1}\Theta}}{1-\cos\varphi} (-d\varphi + \sqrt{-1} \sin\varphi d\Theta) ,$$

$$ds^2 = (1 - \cos\varphi)^2 \, dwd\bar{w}$$
$$= \frac{4}{(1+w\bar{w})^2} dwd\bar{w} , \tag{2.2.2}$$

where ds is the Euclidean metric on S^2, i.e.,

$$ds^2 = d\varphi^2 + \sin^2\varphi d\Theta^2 .$$

We set

$$e_\Theta = (-\sin\Theta, \cos\Theta, 0), \quad e_\varphi = (\cos\varphi\cos\Theta, \cos\varphi\sin\Theta, -\sin\varphi) .$$

These are the unit tangent vectors to the polar coordinate curves. Then e_Θ, e_φ constitute a tangent orthogonal frame on S^2. To them correspond complex vectors w_Θ, w_φ by means of (2.2.1) and we have a relation $w_\varphi = \sqrt{-1}\, w_\Theta$. We, therefore, can define a comlex structure J by $J(e_\Theta)=e_\varphi$ and thus the tangent bundle of S^2 becomes a complex line bundle. Choosing e_Θ a base for it, we have

$$De_\Theta = -\sqrt{-1} \cos\varphi \, d\Theta e_\Theta .$$

Thus the connection is given by

$$\omega = -\sqrt{-1} \cos\varphi d\Theta .$$

The curvature form is

$$F = \sqrt{-1}\ \sin\varphi d\varphi_\wedge d\Theta = \frac{2dw_\wedge dw}{(1+ww)^2}.$$

We write $R_+ = \{r \in R \mid r > 0\}$ and define a map

$$\iota:\ R_+ \times S^2 \xrightarrow{\sim} E^3 - 0,$$

by $\iota(r, (x',y',z')) = (rx',ry',rz')$. By use of this map we can define a connection form on E^3-0 by $(\iota^{-1})^*(\omega)$.

2.3 <u>Universal Connections</u>

Universal bundle connection

The complex projective plane $CP(1)$ has the complex lines through the origin in complex Euclidean 2-space E^2_C as its underlying set. Let $L \in CP(1)$. L is determined by homogeneous coordinates z_0, z_1. Supposing $z_0 \neq 0$, we set $z = z_1/z_0$. Then we can write

$$z = \tan\varphi e^{\sqrt{-1}\Theta} \qquad \text{for some } \varphi,\ \Theta \in R.$$

The line L is generated by vector $(1,z)$, whence by $\xi=(\cos\varphi, \sin\varphi e^{\sqrt{-1}\Theta})$. ξ is a unit vector. We take another unit vector $\eta=(-\sin\varphi, \cos\varphi e^{\sqrt{-1}\Theta})$ orthogonal to ξ. Then we have

$$d\xi = \sqrt{-1}\ \sin^2\varphi d\Theta\xi + (d\varphi + \sqrt{-1}\ \cos\varphi\sin\varphi d\Theta)\eta .$$

We define the absolute differential of ξ as

$$D\xi = \sqrt{-1}\ \sin^2\varphi d\Theta .$$

Hence the connection form ω is $\sqrt{-1}\ \sin^2\varphi d\Theta$. Since $\bar{z}/z = e^{-2\sqrt{-1}\Theta}$, we get

$$d\Theta = \frac{\bar{z}dz-zd\bar{z}}{2\sqrt{-1}z\bar{z}} .$$

We therefore have

$$\omega = (1 - \frac{1}{1+z\bar{z}})\ \frac{\bar{z}dz-\bar{z}dz}{2z\bar{z}} = \frac{zd\bar{z}-zd\bar{z}}{2(1+z\bar{z})} .$$

Hence the curvature form is

$$F = \frac{-dz\wedge d\bar{z}}{(1+z\bar{z})^2} .$$

CP(1) is of course nothing but the Riemann sphere. But this connection defines a different connection on E^3 from the one obtained by use of the map ι .

2.4 One More Connection on E^3

Trivial bundle connection

Let us introduce another connection into E^3 by the following connection matrix:

$$\omega = \frac{1}{r^2}\begin{pmatrix} 0 & xdy-ydx & xdz-zdx \\ ydx-xdy & 0 & ydz-zdy \\ zdx-xdz & zdy-ydz & 0 \end{pmatrix},$$

where $r=\sqrt{x^2+y^2+z^2}$. On the other hand let us define a section Φ of the trivial bundle $E^3 \times so(3)$ over the base E^3, except for the origin by the following expression

$$\Phi = \frac{1}{r}\begin{pmatrix} 0 & z & -y \\ -z & 0 & x \\ y & -x & 0 \end{pmatrix}.$$

Then we have

$$\begin{aligned} D\Phi &= d\phi - \Phi\omega + \omega\Phi \\ &= \frac{1}{r}\begin{pmatrix} 0 & dz & -dy \\ -dz & 0 & dx \\ dy & -dx & 0 \end{pmatrix} - \frac{dr}{r^2}\begin{pmatrix} 0 & z & -y \\ -z & 0 & x \\ y & -x & 0 \end{pmatrix} \\ &\quad - \frac{1}{r^3}\begin{pmatrix} 0 & r^2dz-zrdr & -r^2dy+yrdr \\ zrdr-r^2dz & 0 & r^2dx-xrdr \\ r^2dy-yrdr & xrdr-r^2dx & 0 \end{pmatrix} \\ &= 0 . \end{aligned} \qquad (2.4.1)$$

On the other hand we see the curvature form has this form:

$$F = \frac{1}{r^3}(xdy\wedge dz + ydz\wedge dx + zdx\wedge dy)\Phi .$$

This implies $*F=\frac{dr}{r^2}\Phi$. Hence it follows from (2.4.1) that $D*F=0$. As is known, the Killing form $B(A,A')$ of $\mathfrak{so}(3)$ $(A,A'\in \mathfrak{so}(3))$ is given by $T_r(AA')$. Since $B(\Phi,\Phi)=-2$, we have

$$B(F,\Phi) = \frac{-2}{r^3}(xdy\wedge dz + ydz\wedge dx + zdx\wedge dy) .$$

We see

$$dB(F,\Phi) = 0 .$$

Thus $B(F,\Phi)$ gives rise to a magnetic field on E^3, whose components H_x, H_y, H_z are

$$-\frac{2x}{r^3}, -\frac{2y}{r^3}, -\frac{2z}{r^3},$$

respectively. (cf. [3].)

2.5 Connections in general [4],[5]

We consider a differentiable manifold X and a principal H-bundle P over X with projection π where H is supposed to be a Lie group. Suppose $X=\bigcup_{\lambda\in\Lambda} U_\lambda$ where Λ is a set of indices and that P is trivial on U_λ for each $\lambda\in\Lambda$. The local triviality is given by the following φ_λ

$$\pi^{-1}(U_\lambda)\ni \sigma_\lambda(x)h \xrightarrow[\varphi_\lambda]{\sim} (x,h)\in U_\lambda \times H .$$

Let us suppose $\sigma_\lambda(x)h_\lambda=\sigma_\mu(x)h_\mu$ for $x\in U_\lambda\cap U_\mu$. Then we define the transition function by

$$g_{\mu\lambda}h_\lambda = h_\mu .$$

Let us denote by $\mathfrak{h}$ the Lie algebra of H, which is identical to the tangent space of H at the identity. Take $A\in\mathfrak{h}$. Then $A':h \longrightarrow hA$ is a vector field on H where h ranges over H. We use A instead of A' in what follows. We also use the same letter or symbol for a map and its differential. Define a vector field A^* on P by the equality:

$$A^*|\pi^{-1}(x) = (\varphi_\lambda|\pi^{-1}(x)^{-1}(A)) .$$

Then we see

$$A^*h = (\mathrm{Ad}(h^{-1})A)^* .$$

A connection form ω on P is a $\mathfrak{h}$-valued differential 1-form on P such that

$$\omega(A^*) = A \text{ and } \omega(Xh) = \mathrm{Ad}(h^{-1})(\omega(X)) ,$$

where X is a tangent vector to P.

Let us consider a representation $\chi: H \longrightarrow GL(E_0)$, where E_0 is a finite dimensional vector space. H acts on $P\times E_0$ by $(\xi,v) \longrightarrow (\xi h,\chi(h)^{-1}v)$. Then we can define a vector bundle E by $P\times E_0/H$, called "associated vector bundle to P" by the representation χ. If we take the adjoint representation of H, we get the adjoint bundle ad P of P.

Let Y be a vector field on $U_\lambda \subset X$. We define a $\mathfrak{h}$-valued differential 1-form ω_λ on U_λ as follows:

$$\omega_\lambda(\gamma) = \omega(\sigma_\lambda(\gamma)) .$$

Then,

$$\begin{aligned}
\omega_\mu(\gamma) &= \omega(\gamma\sigma_\mu) = \omega(\gamma(\sigma_\lambda\cdot g_{\lambda\mu})) \\
&= \omega(\gamma\sigma_\lambda\cdot g_{\lambda\mu} + \sigma_\mu g_{\lambda\mu}^{-1}\gamma g_{\lambda\mu}) \\
&= \mathrm{Ad}(g_{\lambda\mu}^{-1})\omega_\lambda(\gamma) + \omega(\sigma_\mu(g_{\lambda\mu}^{-1}\cdot\gamma g_{\lambda\mu})) ,
\end{aligned}$$

i.e.,

$$\omega_\mu(\gamma) = \mathrm{Ad}(g_{\lambda\mu}^{-1})\omega_\lambda(\gamma) + g_{\lambda\mu}^{-1}\cdot\gamma g_{\lambda\mu} . \tag{2.5.1}$$

Giving a connection form is equivalent to giving $\{\omega_\lambda\}_{\lambda\in\Lambda}$ satisfying (2.5.1).

Take a section $\rho\in\Gamma(U_\lambda,E)$, which may be considered as an E_0-valued function. Then the absolute differentiation is given by

$$D_\gamma\rho = \gamma\rho + \chi(\omega_\lambda(\gamma))\rho . \tag{2.5.2}$$

Take the adjoint representation for χ. Then we call each $\chi(\omega_\lambda)$ a connection form for ad P.

3. PROOF

3.1 Stereographic Projection

Let S^n be the unit n-sphere with the origin as its center in E^{n+1}, Euclidean (n+1)-space. Then S^n is identicl to the homogeneous space $SO(n+1)/SO(n)$. We write P for the principal bundle $SO(n+1) \longrightarrow S^n$. Consider the associated vector bundle to P by the adjoint representation of $SO(n)$ (which is the structural group of P) and denote it by ad P. Let $X = \bigcup_{\lambda\in\Lambda} U_\lambda$ be an appropriate open covering of X. A connection for ad P is given by a system of $\mathfrak{so}(n)$-valued 1-forms ω_λ on U_λ satisfying (2.5.1). The curvature form $F_\lambda = dw_\lambda + w_\lambda \wedge w_\lambda$ (and also $*F_\lambda$) may be considered as a section of ad $P \otimes \Lambda T^*(X)$. The absolute differentiations of F and *F are given by (2.5.2), i.e. we have

$$DF = dF + \omega\wedge F - F\wedge\omega \quad (=0) ,$$

$$D{*F} = d{*F} + \omega\wedge{*F} - (-1)^{n-2}{*F}\wedge\omega .$$

But we must add that connections dealt with in this Section are defined only on $X = S^n - (0,\cdots,0,1)$.

Let us suppose that $(y_1,\cdots,y_{n+1}) \in E^{n+1}$ be on S^n. Then, setting

$$x_1 = \frac{y_1}{1-y_{n+1}} ,\cdots, x_n = \frac{y_n}{1-y_{n+1}} ,$$

we have the stereographic projection of X onto E^n. We define $ds_1^2 = dx_1^2 + \cdots + dx_n^2$ on E^n and $ds_1^2 = dy_1^2 + \cdots + dy_{n+1}^2$ on X. Then, directly from (2.2.2), we have

$$ds_2^2 = \frac{4}{(1+r^2)^2}\, ds_1^2 ,$$

where $r = \sqrt{x_1^2 + \cdots + x_n^2}$. Through the stereographic projection we can identify $so(n)$-valued forms on E^n with those on X. We write an absolute differentiation on E^n and the corresponding differentiation on X as D at the same time.

Denoting the *-operator with respect to ds_2 by $*_2$ (* indicates that of ds_1), we have

$$D*_2 = d\left(\frac{2}{1+r^2}\right)^{n-2k} \Lambda * + \left(\frac{2}{1+r^2}\right)^{n-2k} D* \tag{3.1.1}$$

for k-forms.

3.2 Monopoles on E^n

Let us define a connection form ω on Euclidean n-space in the following way.

$$\omega = \frac{1}{r^2}\begin{pmatrix} 0 & x_1dx_2-x_2dx_1 & \cdots & x_1dx_n-x_ndx_1 \\ x_2dx_1-x_1dx_2 & 0 & \cdots & x_2dx_n-x_ndx_2 \\ \cdot & \cdot & \cdot & \cdot \\ x_ndx_1-x_1dx_n & \cdot & \cdot & 0 \end{pmatrix}. \tag{3.2.1}$$

We write the curvature form of ω as F. We first note the following equalities.

$$rdr = x_1dx_1 + \cdots + x_ndx_n .$$

$$\begin{aligned}&(x_1dx_1 + \cdots + x_ndx_n) \wedge (x_idx_j - x_jdx_i) \\ &= \Big(\sum_{k\neq i,j} x_kdx_k + x_idx_i + x_jdx_j\Big) \wedge (x_idx_j - x_jdx_i) \\ &= - \sum x_k(x_idx_j\wedge dx_k + x_jdx_k\wedge dx_i + x_kdx_i\wedge dx_j) + r^2dx_i\wedge dx_j .\end{aligned}$$

On the other hand

$$\frac{1}{2}\, d\omega = - \frac{dr}{r} \wedge\omega + \frac{1}{r^2}\begin{pmatrix} 0 & dx_1\wedge dx_2 & \cdots & dx_1\wedge dx_n \\ dx_2\wedge dx_1 & 0 & \cdots & dx_2\wedge dx_n \\ \cdot & \cdot & \cdot & \cdot \\ dx_n\wedge dx_1 & \cdot & \cdot & 0 \end{pmatrix},$$

where $r=\sqrt{x_1^2+\cdots+x_n^2}$. We write A in what follow instead of

$$\begin{pmatrix} 0 & dx_1\wedge dx_2 & \cdots & dx_1\wedge dx_n \\ dx_2\wedge dx_1 & 0 & \cdots & dx_2\wedge dx_n \\ \cdot & \cdot & \cdot & \cdot \quad \cdot \\ dx_n\wedge dx_1 & \cdot & \cdot \ \cdot \ \cdot & \cdot \quad 0 \end{pmatrix}.$$

Let us denote the (i,j)-element of $r^4\omega\wedge\omega$ by B_{ij}. Then we have

$$B_{ij} = \sum_{k=1}^{n}(x_i dx_k - x_k dx_i)\wedge(x_k dx_j - x_j dx_k)$$

$$= \sum_{k=1}^{n} x_k dx_k\wedge(x_i dx_j - x_j dx_i) - r^2 dx_i\wedge dx_j . \qquad (3.2.2)$$

Hence we can see

$$\omega\wedge\omega = \frac{dr}{r}\wedge\omega - \frac{1}{r^2}A = -\frac{1}{2}d\omega .$$

Thus we find the curvature form $F=d\omega+\omega\wedge\omega$ is equal to $\frac{1}{2}d\omega$. It follows from (3.2.2) that we have

$$\frac{dr}{r}\wedge\omega = \frac{1}{r^2}(x_1dx_1+\cdots+x_ndx_n)\wedge\omega$$

$$= -\frac{1}{r^4}C + \frac{1}{r^2}A ,$$

where C is the matrix the (i,j)-element of which is

$$-B_{ij} = \sum_{k\neq i,j} x_k(x_i dx_j\wedge dx_k + x_j dx_k\wedge dx_i + x_k dx_i\wedge dx_j) .$$

Finally we can get

$$F = \frac{1}{r^4}C .$$

We, therefore, have

$$*F = \Big(\frac{1}{r^4}\sum_{k\neq i,j}\pm x_k(x_i dx_i + x_j dx_j + x_k dx_k)\wedge \underbrace{dx_1\wedge\cdots\wedge dx_n}_{i,j,k:\text{omitted}}\Big)_{i,j}$$

$$= \frac{dr}{r^3}\wedge\Big(\sum_{k\neq i,j}\pm x_k \underbrace{dx_1\wedge\cdots\wedge dx_n}_{i,j,k:\text{omitted}}\Big)_{i,j} , \qquad (3.2.3)$$

where $\pm$ should be decided so that

$$dx_i \wedge dx_j \wedge dx_k \wedge \underbrace{dx_1 \wedge \cdots \wedge dx_n}_{i,j,k:\text{omitted}} = \pm dx_1 \wedge \cdots \wedge dx_n \,.$$

Further we have

$$d{*}F = -\,(n-2)\,\frac{dr}{r^3} \wedge (\pm \underbrace{dx_1 \wedge \cdots \wedge dx_n}_{i,j:\text{omitted}})_{i,j}\,, \tag{3.2.4}$$

where $\pm$ should be decided so that

$$dx_i \wedge dx_j \wedge \underbrace{dx_1 \wedge \cdots \wedge dx_n}_{i,j:\text{omitted}} = \pm dx_1 \wedge \cdots \wedge dx_n \,.$$

3.3 The Case n=4

$$*F = \frac{dr}{r^3} \wedge \begin{pmatrix} 0 & x_3dx_4-x_4dx_3 & x_4dx_2-x_2dx_4 & x_2dx_3-x_3dx_2 \\ x_4dx_3-x_3dx_4 & 0 & x_1dx_4-x_4dx_1 & x_3dx_1-x_1dx_3 \\ x_2dx_4-x_4dx_2 & x_4dx_1-x_1dx_4 & 0 & x_1dx_2-x_2dx_1 \\ x_3dx_2-x_2dx_3 & x_1dx_3-x_3dx_1 & x_2dx_1-x_1dx_2 & 0 \end{pmatrix}.$$

It sufficies to calculate only the upper triangular parts of $\omega \wedge {*F}$ and ${*F} \wedge \omega$ and they turn out to have the following forms as shown by an easy calculation:

$$\omega \wedge {*F} = -\frac{2dr}{r^5} \wedge$$

$$\begin{pmatrix} * & (x_1dx_3-x_3dx_1)\wedge(x_4dx_1-x_1dx_4) & (x_1dx_2-x_2dx_1)\wedge(x_1dx_4-x_4dx_1) & (x_1dx_2-x_2dx_1)\wedge(x_3dx_1-x_1dx_3) \\ & * & (x_2dx_1-x_1dx_2)\wedge(x_4dx_2-x_2dx_4) & (x_2dx_1-x_1dx_2)\wedge(x_2dx_3-x_3dx_2) \\ & & * & (x_3dx_1-x_1dx_3)\wedge(x_2dx_3-x_3dx_2) \\ & & & * \end{pmatrix},$$

$${*F} \wedge \omega = -\frac{2dr}{r^5} \wedge$$

$$\begin{pmatrix} * & (x_4dx_2-x_2dx_4)\wedge(x_3dx_2-x_2dx_3) & (x_3dx_4-x_4dx_3)\wedge(x_2dx_3-x_3dx_2) & (x_3dx_4-x_4dx_3)\wedge(x_2dx_4-x_4dx_2) \\ & * & (x_4dx_3-x_3dx_4)\wedge(x_1dx_3-x_3dx_1) & (x_4dx_3-x_3dx_4)\wedge(x_1dx_4-x_4dx_1) \\ & & * & (x_2dx_4-x_4dx_2)\wedge(x_1dx_4-x_4dx_1) \\ & & & * \end{pmatrix}.$$

The (1,2)-element of $\omega\wedge{*F} - {*F}\wedge\omega$

$$= \frac{2dr}{5}\wedge\{(x_1^2+x_2^2)dx_3\wedge dx_4-(x_1dx_1+x_2dx_2)\wedge(x_3dx_4-x_4dx_3)\}$$

$$= \frac{2dr}{r^5}\wedge\{(x_1^2+x_2^2+x_3^2+x_4^2)dx_3\wedge dx_4-(x_1dx_1+x_2dx_2+x_3dx_3+x_4dx_4)\wedge(x_3dx_4-x_4dx_3)\}$$

$$= \frac{2dr}{r^3}\wedge dx_3\wedge dx_4 \ .$$

On the other hand we have

$$d{*F} = -\frac{2dr}{r^3}\wedge\begin{pmatrix} 0 & dx_3\wedge dx_4 & dx_4\wedge dx_2 & dx_2\wedge dx_3 \\ dx_4\wedge dx_3 & 0 & dx_1\wedge dx_4 & dx_3\wedge dx_1 \\ dx_2\wedge dx_4 & dx_4\wedge dx_1 & 0 & dx_1\wedge dx_2 \\ dx_3\wedge dx_2 & dx_1\wedge dx_3 & dx_2\wedge dx_1 & 0 \end{pmatrix} .$$

We, therefore, can conclude

$$d{*F} + \omega\wedge{*F} - {*F}\wedge\omega = 0 \ .$$

This is the proof of our assertion for the 4-dimension.

3.4 Calculation

In this Section we show that

$$D{*F} = d{*F} + \omega\wedge{*F} - (-1)^n{*F}\wedge\omega = 0 \qquad (3.4.1)$$

holds for an arbitrary dimension n.

3.4.1 Step I

$$\omega\wedge{*F} = \frac{dr}{r^5}\wedge\Big(\sum_{h\neq i,j}(x_idx_h-x_hdx_i)\wedge\sum_{k\neq j,h}(\pm x_k\underbrace{dx_1\wedge\cdots\wedge dx_n}_{h,j,k})\Big). \qquad (3.4.1.1)$$

($\underbrace{\quad}_{h,j,k}$ indicates that the factors dx_h, dx_j, dx_k should be omitted. We will follow this convention from now on). We calculate the (i,j)-element of (3.4.1.1)

$$\sum_{h\neq i,j}x_idx_h\wedge\sum_{k\neq j,h}\pm x_k\underbrace{dx_1\wedge\cdots\wedge dx_n}_{h,j,k} = \sum_{\substack{h\neq i,j\\ k\neq j,h}}\pm x_ix_k\underbrace{dx_1\wedge\cdots\wedge dx_n}_{j,k}$$

$$= -(n-2)(\pm x_i^2 \underbrace{dx_1\wedge\cdots\wedge dx_n}_{i,j}) + (n-3)\sum_{k\neq i,j} \pm x_i x_k \underbrace{dx_1\wedge\cdots\wedge dx_n}_{j,k} .$$

$$\sum_{h\neq i,j} x_h dx_i \wedge \sum_{k\neq j,h} \pm x_k \underbrace{dx_1\wedge\cdots\wedge dx_n}_{h,j,k} = -\sum_{h\neq i,j} \pm x_h (x_i dx_i)\wedge \underbrace{dx_1\wedge\cdots\wedge dx_n}_{i,j,h} .$$

$$\sum_{k\neq i,j} \pm x_i x_k \underbrace{dx_1\wedge\cdots\wedge dx_n}_{j,k} = \sum \pm x_k (x_i dx_i)\wedge \underbrace{dx_1\wedge\cdots\wedge dx_n}_{i,j,k} .$$

Hence we have

$$\omega\wedge *F = -\frac{dr}{r^5}\,(*(n-2)(\pm x_i^2 \underbrace{dx_1\wedge\cdots\wedge dx_n}_{i,j}) + (n-2)\sum \pm x_k (x_i dx_i)\wedge \underbrace{dx_1\wedge\cdots\wedge dx_n}_{i,j,k})_{i,j} .$$

(3.4.1.2)

3.4.2 <u>Step II</u>

$$*F\wedge\omega = \frac{dr}{r^5}\wedge\Big(\sum_{h\neq i,j}\ \sum_{\ell\neq i,h} (\pm x_\ell \underbrace{dx_1\wedge\cdots\wedge dx_n}_{i,h,\ell})\ (x_h dx_j - x_j dx_h)\Big) .$$

$$\sum_{h\neq i,j}\ \sum_{\ell\neq i,h} (\pm x_\ell \underbrace{dx_1\wedge\cdots\wedge dx_n}_{i,h,\ell} \wedge (x_h dx_j)) = \sum_{h\neq i,j} \pm(-1)^{n-3} x_h (x_j dx_j)\wedge \underbrace{dx_1\wedge\cdots\wedge dx_n}_{i,h,j}$$

$$= (-1)^n \sum_{k\neq i,j} \pm x_k (x_j dx_j)\wedge \underbrace{dx_1\wedge\cdots\wedge dx_n}_{i,j,k} .$$

$$\sum_{h\neq i,j}\ \sum_{\ell\neq i,h} \pm x_\ell \underbrace{dx_1\wedge\cdots\wedge dx_n}_{i,h,\ell} \wedge (x_j dx_h)$$

$$= \sum_{h\neq i,j}\ \sum_{\ell\neq i,h} \pm(-1)^{n-3} x_\ell x_j \underbrace{dx_1\wedge\cdots\wedge dx_n}_{i,\ell}$$

$$= (-1)^n (n-2)(\pm x_j^2 \underbrace{dx_1\wedge\cdots\wedge dx}_{i,\ j}) + (-1)^n (n-3) \sum_{k\neq i,j} \pm x_k x_j \underbrace{dx_1\wedge\cdots\wedge dx_n}_{i,k}$$

$$= (-1)^n\{\pm(n-2)x_j^2 \underbrace{dx_1\wedge\cdots\wedge dx_n}_{i,j} - (n-3)\sum_{k\neq i,j} \pm x_k(x_j dx_j)\wedge \underbrace{dx_1\wedge\cdots\wedge dx_n}_{i,j,k}\} .$$

We therefore obtain

$$*F\wedge\omega = -(-1)^n \frac{dr}{r^5}\wedge(\pm(n-2)x_j^2 \underbrace{dx_1\wedge\cdots\wedge dx_n}_{i,j} - (n-2)\sum_{k\neq i,j} \pm x_k(x_j dx_j)\wedge \underbrace{dx_1\wedge\cdots\wedge dx_n}_{i,j,k})_{i,j} \cdot \qquad (3.4.2.1)$$

Using (3.4.1.2) and (3.4.2.1), we can see

$$\omega\wedge *F-(-1)^n *F\wedge\omega$$

$$= -(n-2)\frac{dr}{r^5}\wedge -(\pm(x_i^2+x_j^2)\underbrace{dx_1\wedge\cdots\wedge dx_n}_{i,j} + \sum_{k\neq i,j} \pm x_k(x_i dx_i + x_j dx_j)\wedge \underbrace{dx_1\wedge\cdots\wedge dx_n}_{i,j,k})$$

$$= -(n-2)\frac{dr}{r^5}\wedge(-(\pm r^2 \underbrace{dx_1\wedge\cdots\wedge dx_n}_{i,j}) + \sum_{k\neq i,j} \pm x_k(x_i dx_i + x_j dx_j + x_k dx_k)\wedge \underbrace{dx_1\wedge\cdots\wedge dx_n}_{i,j,k})$$

$$= \pm(n-2)\frac{dr}{r^3}\wedge(\underbrace{dx_1\wedge\cdots\wedge dx_n}_{i,j} - (n-2)\frac{dr}{r^5}\wedge\sum_{k\neq i,j} \pm x_k(x_1 dx_1 + \cdots + x_n dx_n)\wedge \underbrace{dx_1\wedge\cdots\wedge dx_n}_{i,j,k})$$

$$= \pm(n-2)\frac{dr}{r^3}\wedge(\underbrace{dx_1\wedge\cdots\wedge dx_n}_{i,j}) . \qquad (3.4.2.2)$$

From (3.2.4) and (3.4.2.2) we can conclude $D*F=0$.

3.5 Monopoles on S^n

In the previous Section we showed the existence of a monopole with one-point singularity on E^n. Then the Yang-Mills equation on $\tilde{X}=S^n-\pm(0,\cdots,0,1)$, $D*_2F=0$, follows from (3.1.1) and (3.2.2) in such a way that

$$D*_2F = \left(\frac{2}{1+r^2}\right)^{n-4} D*F - (n-4)\left(\frac{2}{1+r^2}\right)^{n-3} rdr\wedge *F = 0$$

APPENDIX 1

Electro-magnetic fields. Let L be a complex line bundle over a real differential manifold X. Let $\{U_\lambda\}$ be a system of coordinate neighbourhoods indexed by $\lambda \in \Lambda$ and $\{g_{\lambda\mu}\}$ a system of transition functions (we can assume $|g_{\lambda\mu}| = 1$.) A (metric preserving) connection is given by 1-forms ω_λ on U_λ satisfying

$$\omega_\mu = \omega_\lambda + d \log g_{\lambda\mu}, \quad \omega_\lambda + \bar{\omega}_\lambda = 0 .$$

We write $F = -\sqrt{-1}\, d\omega_\lambda$. Then, of course

$$dF = 0 . \tag{1}$$

Suppose that X is Minkowski space-time. Set

$$F = (E_x dx + E_y dy + E_z dz)\wedge dt + (H_x dy\wedge dz + H_y dz\wedge dx + H_z dx\wedge dy) .$$

Then $E = (E_x, E_y, E_z)$ and $H = (H_x, H_y, H_z)$ give rise to an electromagnetic field with electric charge density and electric current density. In addition to (1),

$$d{*}F = 0 \tag{2}$$

constitutes Maxwell's equations in the vacuum.

APPENDIX 2

Of the connection ω defined by (3.2.1) the following facts are characteristic.

(1) $\frac{x}{r}$ gives rise to a parallel vector field.

(2) $D\, dx = r\, dr\ \ dx$.

(3) It preserves the metric of E^n.

Let us denote by $\tilde{\omega}$ the pullback of ω by the inclusion $S^{n-1} \subset E^n$ and consider a map $\pi : E^n - 0 \longrightarrow S^{n-1}$ which is defined by $\pi(x) = \frac{x}{r}$ where $x \in E^n - 0$. Then we have $\pi^*(\tilde{\omega}) = \omega$. $\tilde{\omega}$ defines a connection on the product bundle $S^{n-1} \times E^n$ over S^{n-1}. Naturally $T(S^{n-1}) \subset S^{n-1} \times E^n$. It follows from (1) and (3) that the tangent spaces of S^{n-1} are parallel to one another. We write ω_S for the induced connection of $\tilde{\omega}$ on $T(S^{n-1})$. It follows from (2) and (3) that ω_S coincides with the usual Levi-Civita connection on S^{n-1}. With these settings we can see that (3.4.1) is an immediate consequence of the fact that ω_S satisfies the Yang-Mills equations on S^{n-1}. The connection ω belongs to the category of Uhlenbeck's examples [6].

For the Minkowski space with the metric $ds^2 = dx_1^2 + \cdots + dx_n{}^2 - dt^2$ we can find a similar connection. Its precise form is as follows.

$$\omega_M = \frac{1}{r^2}\begin{pmatrix} 0 & x_1dx - x_2dx & \cdots & x_1dt - t\,dx_1 \\ \vdots & \vdots & & \vdots \\ x_ndx_1 - x_1dx_n & \cdots & 0 & x_ndt - t\,dx_n \\ x_1dt - tdx_1 & \cdots & x_ndt - tdx_n & 0 \end{pmatrix},$$

where $r^2 = x_1{}^2 + \cdots + x_n{}^2 - t^2$.

This satisfies the Yang-Mills equations.

The unitarian case or the quaternionic case is a little more complicated and seems much more interesting. We would like to talk about it somewhere else.

REFERENCES

[1] Jaffe, A. and Taubes, C.H., Vortices and Monopoles, Boston, Birkhäuser, 1980.

[2] Goddard, P. and Olive, D.I., New Developments in the Theory of Magnetic Monopoles, CERN Preprint.

[3] t'Hooft, G.: Magnetic monopoles in unified Gauge theories, Nucl. Phy., B79, 276-284 (1974).

[4] Kobayashi, S. and Nomizu, K., Foundations of Differential Geometry I, Interscience Publishers, New York, 1963.

[5] Koszul, J.L., Lectures on Fibre Bundles and Differntial Geometry, Tata Institute, 1960.

[6] Sibner, K.M.: The isolated point singularity problem for the coupled Yang-Mills equations in higher dimensions, Math. Ann., 271, 125-131 (1985).

Proc. Prospects
of Math. Sci.
World Sci. Pub.
97-125, 1988

INTEGRATION AND NUMBER THEORY

JEAN-LOUP MAUCLAIRE
L'U.E.R. de Mathématique et Informatique
Université Paris VII
2, Place Jussieu
75251 Paris
France

0. INTRODUCTION

The purpose of this paper is to present some aspects of the use of integration for the study of arithmetical functions and their generalizations.

1. SOME HISTORICAL COMMENTS

1-a. It is generally admitted that the starting of the probabilistic number theory was settled, in the thirties, with the name of Turán, Erdös and so on. We shall not discuss this assertion, although some titles of Cesaro's works such as "Probabilité de certains faits arithmétiques", "Eventualité de la division arithmétique", prove that the idea of probability in Number Theory can be traced before 1889 [1],[2]. To understand the development of the so-called probabilistic Number Theory, we have to fix an origin, and it would be 1930.

1-b. Although the famous work of Steinhaus "Les probabilités dénombrables et leurs rapports à la théorie de la mesure" appeared in 1923 [3], it is Kolmogorov's monograph "Grundbegriffe der Wahrscheinlichkeitsrechnung " [4] appearing in 1933 that is considered to have firmly established the foundations of probability theory. The study of laws of the sum of independent random variables was then considerably developed, and during the years 1936 to 1939, Mark Kac got some of his most famous results relating to this field. The encounter with Paul Erdös dealing with the theory of arithmetical additive functions led to the famous Erdös-Kac Theorem on the Gaussian law of errors in the theory of additive functions, in 1939 [5]. The same year, the viewpoint of statistical independence allowed to get the Erdös-Wintner Theorem [10] which gives necessary and sufficient conditions for the existence of the distribution function of an additive arithmetical function. To summarize the underlying ideas, it can be said that:

* if p and q are distinct prime numbers, then the divisibility of an integer by p and the divisibility of an integer by q are independent events.
* given a prime p, we denote by $V_p(n)$ the highest power of p which divides an integer n. Now, consider a sequence of integers N_k, $k \in \mathbb{N}$, such that for any prime p, $\lim_{k \to +\infty} V_p(N_k) = +\infty$, and that $V_p(N_k)$ is a non-decreasing function in k. The basic idea of the Erdös' truncation method is that the set of positive integers is the limit of the set of the divisors of N_k, as $k \to +\infty$, and a consequence is as follows: if we can solve a problem for the set of divisors of N_k for any k, the solution of the problem on the limit set is (probably) the "limit of the solution on N_k", as $k \to +\infty$. Later, during the fifties, this viewpoint was adopted and extended by Kubilius [6] and this was possible by, first, obtaining the famous inequality on arithmetical additive functions after his name, and secondly, a better use of finite probability models. (For more details, see [7],[6]).

1-c. In the early forties, another topic was investigated, in connection with harmonic analysis and Bohr theory of almost periodic functions on real line: it was the study of the almost-periodicity of arithmetical functions, and some attempts have been done to give a general theory, by A. Wintner [8] in 1943 and J. Delsarte [9] in 1945, independently. The works of P. Erdös, M. Kac, R. van Kampen, A. Wintner [11], [12] in this field were sufficient to show that there was a strong link between the almost-periodicity of arithmetical functions and the possibility to study them applying the truncation method of P. Erdös. Anyway, the meaning of this almost-periodicity was not at all considered and one of the fathers of the duality theory of compact abelian groups, R. van Kampen, studied the almost-periodicity of arithmetical functions [13] without any group theory. It seems that one of the first to study this problem was E.V. Novoselov [14],[15], [16]; before him, some attempts had been done, for instance by F.I. Mautner [17], but it was not really systematic; (Paradoxically, most of the concepts attributed to E.V. Novoselov were rather well-known, and the original part of his work has been almost completely overlooked. Still now, it is frequent to see inaccuracies in the presentation of what he did and a evident misunderstanding). In the sixties, E.V. Novoselov tried to build a theory of limit-periodic sequences, and to apply it to the study of arithmetical functions, using topology and measure theory but not exactly the well-known correspondence between Fourier coefficients of limit-periodic sequences and Fourier transform of an integrable function on a compact group which is the basic idea in the Bohr theory, or rather in the Bohr-Følner theory [18]. In place of this classical method, he used a wonderfully astute construction and since most of the authors who refer to E.V. Novoselov seem to have overlooked it, I shall devote some pages later to explain this aspect of his work. The originality of the attempt was in the systematic viewpoint: it was clear, in the mind of this author, that there was somewhere a general theory, and he tried to settle it on a firm basis.

A little bit before this time, E. Cohen introduced the notion of "almost even functions" on the multiplicative semi-group of the positive integers [19]; this idea leads W. Schwarz and J. Spilker [20] to define a topology on the positive integers, and after the completion of the space, to prove that the Ramanujan sums form a system which is total in the set of continuous functions on the completed space, which is a compact set E, [20] and moreover, that there exists a natural measure μ such that the set of positive integers is μ-uniformly distributed on E. Later, in 1976, this viewpoint was extended by J. Knopfmacher [21], which covers the case of more general semi-groups; this author also introduced the notions which enabled

J. Spilker and W. Schwarz to introduce the notion of B-almost even functions and settle the relation between the space of these functions and the space of μ-integrable functions defined on E [22].

Recently, Z. Kryzius reconsidered the problem and gave some interesting results in this field [23],[24],[25].

2. SOME MATHEMATICAL COMMENTS

This chapter will be devoted to sketch two different viewpoints, the first one is E.V. Novoselov's, and the second one is the viewpoint of W. Schwarz, J. Spilker, J. Knopfmacher. To avoid ambiguity, only the basic ideas will be discussed here.

2-a. Notations.

We shall use the following notations:

$\mathbb{Q}$ is the set of rational numbers.

$\mathbb{Z}$ is the set of rational integers.

$\mathbb{N}$ is the set of non-negative integers.

$\mathbb{N}^*$ is the set of positive integers.

$\mathcal{P}$ is the set of primes.

In general, the symbols $n, m, \cdots$ will denote positive integers, and p will always denote a prime number.

$\mathbb{Z}_p$ is the ring of p-adic integers, and $\mathbb{Z}_p^*$ the units of $\mathbb{Z}_p$.

dm_p is the Haar-measure on the compact group Z_p ;

dm_p^* the Haar-measure on the multiplicative compact group $\mathbb{Z}_p^*$; each of them is normalized by 1.

G (resp. G^*) will denote the product of the $\mathbb{Z}_p$ (resp. $\mathbb{Z}^*$) , i.e. $G = \prod_p \mathbb{Z}_p$, $G^* = \prod_p \mathbb{Z}_p^*$; on G (resp. G^*) is defined the unique normalized Haar-measure dm (resp. dm^*) and $dm = \bigotimes_p dm_p$, $dm^* = \bigotimes_p dm_p^*$.

If t belongs to G , t is written $t = (t_p)_{p\in\mathcal{P}}$, and if $t_p \neq 0$, t_p can be written $t_p = p^{V_p(t)} \cdot t_p^*$, where t_p^* is in $\mathbb{Z}_p^*$, $V_p(t) \in N$, and such a decomposition is unique.

If X is a ν-measured space, $^{\alpha}(X,\nu)$ (resp. $L^{\alpha}(X,\nu)$) denote the set of complex-valued functions on X which are α^{th}-power ν-integrable (resp. the equivalence classes in $L^{\alpha}(X,\nu)$).

If $a : \mathbb{N}^* \to \mathbb{R}$, then $M(a)$ (resp. $\overline{M}(a)$, $\underline{M}(a)$) will denote the expression $\lim_{x\to+\infty} \frac{1}{x} \sum_{n\leq x} a(n)$, if it exists, (resp. $\limsup_{x\to+\infty} \frac{1}{x} \sum_{n\leq x} a(n)$, $\liminf_{x\to+\infty} \frac{1}{x} \sum_{n\leq x} a(n)$). For two topological spaces X and Y, $C(X,Y)$ denote the space of Y-valued continuous functions defined on X.

2-b. <u>The ideas of E.V. Novoselov.</u>

2-b-1. <u>Basic constructions.</u>

On $\mathbb{Z}$, a basis of neighborhood of 0 is defined by the ideals in $\mathbb{Z}$, $d\mathbb{Z}$, $d\in\mathbb{N}$; it is possible to introduce a metric on $\mathbb{Z}$ to define this topology, and the completion of $\mathbb{Z}$ is the ring of p-adic integers $\mathcal{G}$, which is compact. Since it is a compact additive group, there exists a unique normalized Haar measure dx; moreover, if $f\in C(\mathcal{G},\mathbb{C})$, we have:

$$M(f) = \int f(x)dx .$$

A special family of subsets of $\mathbb{N}^*$ is studied, "the bi-measurable sets", (but the present author proved that they can be identified with the Riemann-integrable sets for the considerated measure dx [31]; so, they are not really exceptional). An important work is the very detailed study of $\mathcal{G}$ from algebraic viewpoint, but we shall omit it here. More important point for us is the following results.

$\mathcal{G}$ is isomorphic to G.
dx is identical to dm.
Any x in $\mathcal{G}$ can be written:

$$x = \sum_{j=o}^{+\infty} a_j \cdot j! , \quad 0 \leq a_j \leq j-1 , \quad j\in\mathbb{N}^* .$$

The dual group of G is $\mathbb{Q}/\mathbb{Z}$.

Now, consider the set P of complex periodical sequences; then, the closure of P by the uniform topology can be identified with $C(G,\mathbb{C})$; we call this closure the set of uniform-limit-periodical sequences, say U.L.P.

Now, define the Ramanujan sums C_q , $q \in \mathbb{N}^*$, by:

$$C_q(n) = \sum_{\substack{(h,q)=1 \\ 0<h<q}} e^{2i\pi \frac{h}{q} n}$$

then, the family of the C_q can be extended to a system of continuous functions, still denoted C_q , which is total in the set of complex-valued continuous functions on G which are invariant under G^* ; moreover, this system is orthogonal, i.e., if $q \neq r$,

$$\int C_q C_r dm = 0 .$$

All these things just described above in the work of E.V. Novoselov; but not all of them are due only to him: the polyadic numbers have been defined by van Dantzig [26],[27],[28],[29],[30], who studied them carefully in 1935; the topology on $\mathbf{Z}$ is classical and appears at least in the thesis of J. Tate; the dual group of G has been studied by M. Abe [32] in 1940, and the notion of U.L.P. function is a direct consequence of this study and an application of A. Weil's ideas [33], etc. ... .

The originality of E.V. Novoselov here is the systematical investigation, by use of topology, measure theory, and it is clear that he wanted to apply powerful theories to study arithmetical functions. It was perfectly clear in his mind that the difficulty was to connect functions defined on G and sequences, that means functions defined on $\mathbb{N}^*$, and all the trouble comes from the fact that the image of $\mathbb{N}^*$ in G has a measure equal to zero. So, in the case of U.L.P. sequences, i.e. restriction of elements of $C(G,\mathbb{C})$ to $\mathbb{N}^*$, the uniform distribution of the sequence $\mathbb{N}^*$ in G allows to avoid this difficulty; the introduction of the bimeasurability, this notion due completely to Buck [34], was an attempt to overcome it too, for E.V. Novoselov was unaware of the relation with Riemann integral, it seems. But it was clear that the "good sets of functions" were the $L^\alpha(G,dm)$, because the limit-periodicity of sequences is linked with these spaces, an immediate consequence of the work of H. Bohr and E. Følner [35],[18] , and many arithmetical functions are related to this set of sequences, as it appears in the works of P. Erdös, Hartman, M. Kac, von Kampen, A. Wintner [11],[12],[13],[36],[37],[8]. To establish the relation between limit-periodical sequences and dm-integrable functions on G, E.V. Novoselov developed a perfectly original method, and I shall explain it now; anyway, I want not to detail lengthy the reasons for which such a tool is effective, nor to give all the proofs; the interested reader is referred to my paper [38], in which all these things are treated in a very simplified way.

2-b-2. <u>From limit-periodicity to integrability</u>.

The first point is the following: There exists a sequence N_k, $k \in \mathbb{N}$, such that

$$N_{k+1} > N_k \ , \ \lim_{k\to+\infty} \frac{N_{k+1}}{N_k} = 1 \ , \ \lim_{k\to+\infty} V_p(N_k) = +\infty \quad \text{for any } p \ .$$

Define $R_k(x)$, a sequence of functions from G to $\mathbb{N}^*$, by : $R_k(x)$ is the smallest element x_k of $\mathbb{N}$ such that $x-x_k$ belongs to N_kG.

Now, if f is a function from G to $\mathbb{C}$, it is clear that $f(R_k(x))$, $k \in \mathbb{N}$, is a sequence of continuous functions from G to $\mathbb{C}$, and $f(R_k(x))$ can be identified to a periodic function on $\mathbb{N}^*$, with period N_k.

We define a function space h^r, $r \geq 1$, by:

$$h^r = \{f : G \to \mathbb{C} \mid \lim_{k\to+\infty} \int_G |f(x)-f(R_k(x))|^r \, dm(x) = 0\} \ .$$

So, h^r is a sub-space of $L^r(G,dm)$.

The idea is that, because of the choice of N_k and the fact that a function $f \in h^r$ is defined by its values on $\mathbb{N}^*$, is possible to prove that

$$M(f) \text{ exists and } M(f) = \int_G f(x) \, dm(x) \ .$$

An essential ingredient, very simple, is that, if f is any real sequence,

$$\overline{M}(f) = \limsup_{k\to+\infty} \int_G f(R_k(x))\, dm(x) ,$$

$$\underline{M}(f) = \liminf_{k\to+\infty} \int_G f(R_k(x))\, dm(x) .$$

Now, consider a sequence f such that, for a given $r \geq 1$, $\lim_{k\to+\infty} \overline{M}(|f-f(R_k)|^r)^{1/r} = 0$. Then, $f(R_k(n))$ can be extended in G; the sequence of extensions is a Cauchy sequence in $L^r(G,dm)$ since

$$M(|f(R_k)-f(R_\ell)|^r)^{1/r} = (\int_G |f(R_k(x))-f(R_\ell(x))|^r\, dm(x))^{1/r} .$$

So the limit $f(x)$ exists in $L^r(G,dm)$, and can be chosen in h^r, for $m(N^*) = 0$. But it can be proved that:

A sequence $f(m)$ satisfies $\lim_{k\to+\infty} M(|f-f(R_k(\cdot))|^r)^{1/r} = 0$ if and only if there exists f_n, a sequence of periodical sequences such that

$$\lim_{n\to+\infty} M(|f-f_n|^r)^{1/r} = 0 .$$

__Definition__. Sequences satisfying the second part of this assertion are called B^r-limit-periodical sequences (B comes from Besicovitch), and this set will be denoted by B^r-L-P.

The conclusion is the following:

__Theorem of E.V. Novoselov__. *Any element of* B^r-L-P *can be extended into an element* f *of* h^r, *and to any element* f *of* h^r *can be associated an element of* B^r-L-P. *Moreover, we have*:

$$M(f) = \int_G f(x)\, dm(x) ,$$

$$M(|f-f(R_k(\cdot))|^r)^{1/r} = (\int_G |f(x)-f(R_k(x))|^r dm(x))^{1/r} .$$

2-b-3. __Some remarks on the preceding construction__.

The construction is perfectly defined, and not at all ambiguous; moreover, it can be proved that it does not depend on the choice of a special sequence N_k: the result is the same for any sequence satisfying the required properties. The relation between the probabilistic properties of an element f of h^r and the restriction of the same element f on $\mathbb{N}^*$ is very clear. It seems that the origin of the consideration of the spaces h^r has to be found in the fact that E.V. Novoselov wanted to get results on additive and multiplicative arithmetrical functions: in these two cases, the approximation by the truncation method of Erdös leads to sequences $f(R_k(n))$ in a natural way. But the approach contains a very strong limitation, that I shall explain here.

Consider a sequence of periodical functions on $\mathbb{N}^*$, f_n, such that:

$$\lim_{\substack{n\to+\infty \\ \ell\to+\infty}} M(|f_n - f_\ell|)^r)^{1/r} = 0 .$$

A classical method of Marcinkiewicz leads to the following result[39]: there exists g belonging to B^r-L-P such that

$$\lim_{n\to+\infty} M(|g-f_n|^r)^{1/r} = 0 .$$

Such a result cannot be obtained by the method of this author, and the correspondence between B^r-L-P and $L^r(G,dm)$ is not possible to state in his way of dealing on this topic. So, this will imply some restrictive viewpoint, which leads, at the end, to a rather pointwise approach and measure theory cannot be really powerfully used.

However, only a few of the works of E.V. Novoselov have been presented here, and the reader is referred to the originals to appreciate the genuine ideas of this author.

Now, we shall deal with the basic ideas of the second tendency, well-represented by W. Schwarz, J. Spilker and J. Knopfmacher.

2-c. The Theory of W. Schwarz - J. Spilker - J. Knopfmacher

Here, only the case of $\mathbb{N}^*$ will be considered. Although it is possible to extend it to some semi-groups, we shall not discuss this here; the interested reader may refer to [24],[21],[40].

2-c-1. The main ideas are the following:

An arithmetical function $f : \mathbb{N}^* \to \mathbb{C}$ is "even modulo k" if there exists a k in $\mathbb{N}^*$ such that for any n in $\mathbb{N}^*$, we have $f((n,k)) = f(n)$, where (n,k) denote the greatest common divisor of n and k. The set $\mathcal{B}_k$ of even functions modulo k is a vector space on C and a basis is the set of Ramanujan sums $C_q(n)$ where q divides k [22]. Then, the set of arithmetical functions $\mathcal{D}$ is identified with $\mathcal{D} = \prod_{n\in\mathbb{N}^*} \mathbb{C}_n$, where any $\mathbb{C}_n$ is a copy of $\mathbb{C}$, and the usual topology on $\mathbb{C}$ allows to define a product topology on ; but $\mathcal{B} = \bigcup_{k\in\mathbb{N}^*} \mathcal{B}_k$ is a dense subalgebra of $\mathcal{D}$, and it is possible to prove that the closure of $\mathcal{B}$ for the uniform norm topology can be identified to the set of continuous complex valued functions defined on E, a "compactification of $\mathbb{N}^*$", which can be described in the following way: E is the product of all the E_p, where E_p is the Alexandrof one-point compactification of the discrete set $\{p^m, m\in\mathbb{N}\}$. Since $C(E,\mathbb{C})$ is the closure of $\mathcal{B}$ and any element of $\mathcal{B}$ has an arithmetical mean-value, any element of $C(E,\mathbb{C})$ has an arithmetical mean-value $M(f)$. Riesz representation theorem gives that there exists a complete and regular Borel measure μ such that, if $f\in C(E,\mathbb{C})$, we have $M(f) = \int_E f\, d\mu$. Now, if we define a norm on $C(E,\mathbb{C})$ by $f \to (M(|f|^q))^{1/q}$ for a fixed $q \geqq 1$, it can be proved that $C(E,\mathbb{C})$ can be completed for this norm and the completion can be identified to $L^q(E,d\mu)$.

2-c-2. Some remarks on the preceding construction.

In the preceding construction, a very immediate device allows to simplify the whole. In fact, the idea of the authors is to find a compact set on which the C_q, $q\in\mathbb{N}^*$, can be defined as continuous functions, and then complete $\mathcal{B}$. This is immediate, for if we denote by $p^\alpha \| q$ the expression "p^α divides q, $(p^\alpha | q)$, $p^{\alpha+1}$ does not divide q, $(p^\alpha \nmid q)$", we have:

$$\sum_{d|q} C_d(n) = \sum_{d|q} \sum_{(h,d)=1} e^{2i\pi \frac{h}{d} n} = \sum_{d|q} \sum_{(h,d)=1} e^{2i\pi \frac{h(\frac{q}{d})}{q} n}$$

$$= \sum_{\ell=0}^{q-1} e^{2i\pi \frac{\ell}{q} n} = q\, I_q(n) ,$$

where I_q is defined by $I_q(n) = \begin{cases} 1 & \text{if } q|n \\ 0 & \text{otherwise} \end{cases}$.

Since $I_q(n) = \prod_{p^\alpha \| q} I_{p^\alpha}(n)$, we are looking for a topology on which each I_{p^α} is continuous: only one possibility is allowed, for $I^{-1}_{p^\alpha}\{0\}$ is a projective sequence of subsets of $\mathbb{N}^*$, and so, only the set $\lim\limits_{\alpha\to+\infty} \text{proj}\ \{p^r, r = 0 \text{ to } \alpha\}$ is allowed. This set is identical to E_p, but the measure is defined immediately. Moreover, since we have

$$\{p^r, r = 0 \text{ to } \alpha\} = ((\mathbf{Z}/p^{\alpha+1}\mathbf{Z}) / (\mathbf{Z}/p^{\alpha+1}\mathbf{Z})^* - \{0\}) ,$$

it is possible to "take the limit" and get the fact that the theory developed by these authors in the case of N^* is, except for some details, a part of the theory of Novoselov; the main difference is that they cannot deal with the group G, neither with the spaces h^r, but they have a "correspondence theorem" for the spaces $L^q(E,d\mu)$, i.e. $L^q(G/G^*,d\mu)$, the space of classes of functions of $L^q(G,dm)$ invariant almost-everywhere under G^*;

Anyway, this "correspondence theorem" seems not to give more information on the B^r-L-P arithmetical functions than the method of Novoselov, and one of the reasons is the fact that the approach of the problem via Gelfand theory of Banach Algebras is so general that the exceptional structure of the set of integers is completely disregarded.

Since all these methods appeal, at end, to the truncation method of P. Erdös, we shall follow a different way: in place of trying to find a structure and then embed in it the truncation method, we shall do the contrary: start with the truncation method and find the structure which is naturally deduced.

3. PROJECTIVE MEASURES AND ARITHMETICAL FUNCTIONS

We shall use the same notations as in 2-a.

3-A. Projective Limits of Measures.

3-A-1. From truncation method to probability spaces.

As it has been stated briefly in 1-b , the basic idea of the truncation method is, first to consider a (well-chosen!) sequence N_k in $\mathbb{N}^*$ such that: for any p in P , $V_p(N_k)$ is non-decreasing and $\lim_{k\to+\infty} V_p(N_k) = +\infty$; next solve the problem in $\mathbb{Z}/N_k\,\mathbb{Z}$; then, finally "take the limit: of the solution, $k \to +\infty$. The idea is to use some finite models, and take the limit of the estimates, as it was mentioned in II-c. In our method, we do the contrary: First take infinite models directly, then solve the problem in the probability space associated, and go back to the original problem on integers. Two possibilities occur:

a) if we deal with $\mathbb{N}^*$, and the truncation is done in such a way that the finite spaces are $\mathbb{Z}/N_k\,\mathbb{Z}$, $k = 0$ to $+\infty$, then, the limit space is:

$$\lim_{k\to+\infty} \text{proj}\ \mathbb{Z}/N_k\,\mathbb{Z} = \lim_{k\to+\infty} \text{proj} \prod_p \mathbb{Z}/p^{V_p(N_k)}\mathbb{Z}$$

$$= \prod_p \lim_{k\to+\infty} \text{proj}\ \mathbb{Z}/p^k\,\mathbb{Z} = \prod_p \mathbb{Z}_p = G .$$

We shall use the convention that $N_0 = 1$.

The way in which G is obtained is the starting point of the identification of G with the compact space:

$$\prod_{i=0}^{+\infty} [0,1,\cdots \frac{N_{i+1}}{N_i} - 1] ,$$

for any x in G can be identified with

$$x = a_1(x)+a_2(x)N_1+a_3(x)N_2+\cdots+a_i(x)N_{i-1}+\cdots$$

$$(0 \leq a_i(x) \leq \frac{N_{i+1}}{N_i} - 1 , 0 \leq i)$$

and the natural measure on this compact space can be identified with the Haar-measure on G, via the identification on the Borel sets. So, we get:

In the case a), the truncation method leads to the space (G,dm).

b) If we deal with a problem related to divisibility properties of integers, the truncation methods leads to the study of the spaces

$$\{d \in \mathbb{N}^*, \ d|N_k\},$$

i.e. $\{d \in \mathbb{N}^*, \ d \in ([(\mathbb{Z}/N_k\,\mathbb{Z})\setminus\{0\}] / (\mathbb{Z}/N_k\,\mathbb{Z})^*$

i.e. $\prod_p \{1,p,\cdots,p^{V_p(N_k)}\}$,

and the limit space becomes:

$$\prod_p (\lim_{k\to+\infty} \mathrm{proj}\{1,p,\cdots,p^k\}) = E$$

But it is clear that we may identify E with $\prod_p((\mathbb{Z}_p-\{0\})/\mathbb{Z}_p^*)$, and the associated measure $d\mu$ on E is defined by dm, i.e.:

$$d\mu = \bigotimes_p d\mu_p, \text{ where } \mu_p(p^k) = (1-\frac{1}{p})\frac{1}{p^k}.$$

So, we get:

In the case b), the truncation method leads to the quotient space $((G\setminus\{0\})/G^*, d\mu)$ with the measure $d\mu$ induced by dm.

It must be stressed here that if there exists a relation between the almost-periodicity of arithmetical functions and the possibility to study them by the truncation method, this relation is not at all fortuitous, but rather structural.

3-A-2. Some consequences of the properties of the probability spaces.

The essential property of the structures which appeared in the above presentation of the relation existing between the truncation method and the construction of a probability space, is the fact that they involve projective limits of finite sets. We shall give here some consequences of this fact.

a) The most immediate is the fact that: For any $\alpha \geq 1$, we have:

$$L^{\alpha}(G,dm) = L^{\alpha}(E \times G^{*}, d\mu \otimes dm^{*}) .$$

N.B. It is not true that $\prod_{p} \mathbb{Z}_p = E \times G^{*}$; this statement is attributed to the present author [25] (p.245, line 1, 2), which rejects such an interpretation of the above formula.

b) Consider F in $L^{\alpha}(G,dm)$, $\alpha \geq 1$. Then, we have

$$\int_G F(t+N_k g)dm(g) = f_{k+1}(t) = f_{k+1}\ (t \bmod N_k) ,$$

and f_{k+1} is defined for any t , continuous on G since it is N_k-periodical, and identical to a N_k-periodical function on $\mathbb{N}^{*}$.

i) The martingale convergence theorem gives:

<u>Theorem</u>. $\lim\limits_{k\to+\infty} f_{k+1}(t) = F(t)$ *almost-everywhere and in* $L^{\alpha}(G,dm)$.

N.B. This very special structure allows us to build directly the theory of B^{α}-L-P sequences (defined in 2-2), which we shall explain briefly:

Since $\mathbb{N}^{*}$ is uniformly distributed in G , then, if h is periodical in G , we get, by using obvious notations:

$$\int_G hdm = M(h) .$$

This gives that, if f_{k+1} is a sequence of periodical functions satisfying

$$\lim_{\substack{\ell\to+\infty \\ k\to+\infty}} M(|f_{k+1}-f_{\ell+1}|^{\alpha})^{1/\alpha} = 0 ,$$

there exists F in $L^{\alpha}(G,dm)$ such that

$$\lim_{k\to+\infty} \left(\int_G |F-f_{k+1}|^{\alpha} dm\right)^{1/\alpha} = 0 .$$

Denote by N_k the period of f_{k+1} for any k .

* First, assume that the period of f_k divides the period of f_{k+1}

for any k . Now, define $f(n) = f_{k+1}(n)$ if $N_k < n \leq N_{k+1}$. Then:

$$\lim_{k \to +\infty} (M(|f-f_{k+1}|^{\alpha}))^{1/\alpha} = 0 .$$

So, f and F are associated; practically, this means that they are approximated by the same sequences of periodical functions.

* If the period of N_k does not divide the period of N_{k+1} for some k , we construct a new sequence which satisfies this assumption, in the following way: Let T_i be the period of f_{i+1} .

Define N_k = least common multiple of the T_i , $i = 0$ to k . Define f'_{k+1} by:

f'_{k+1} is the N_k-periodical function on N such that

$$f'_{k+1}(n) = f_{k+1}(n) .$$

Then f'_k will satisfy the conditions: (period of f'_k) divides (period of f'_{k+1}), f'_k is a Cauchy sequence in $L^{\alpha}(G,dm)$. This ends the N.B.

ii) It must be remarked that the Fourier series of $\int_G F(t+N_k g)dm(g)$ has only a finite number of terms, more precisely,

$$\int_G F(t+N_k \cdot g)dm(g) = \sum_{q|N_k} \sum_{(h,q)=1} \hat{F}\left(\frac{h}{q}\right) e^{2i\pi \frac{h}{q} t} ,$$

where $\hat{F}(\frac{h}{q})$ is the Fourier transform of F in $\frac{h}{q}$.

iii) The relation between B^{α}-L-P and $L^{\alpha}(G,dm)$, described in the above N.B., allows to give results by use of measure theory; for instance:

<u>Theorem</u>. *Let* F *in* $^{\alpha}(G,dm)$, f *in* B^{α}-L-P , F *and* f *associated and real.*

Consider the subset of real numbers u *such that* $\lim_{x \to +\infty} \frac{1}{x}$ (cardinal of $\{0 \leq n < x \mid f(n) < u\}$) *exists. Denote by* $\delta(u)$ *that limit.*

Then, if u *is a continuity point of* δ *, we have*

$$\delta(u) = m\{t\in G | F(t) < u\} .$$

Moreover, if I(t) *is defined by* $I(t) = \begin{cases} 1 & \textit{if } F(t) < u \\ 0 & \textit{otherwise} \end{cases}$,

and J(n) *is defined by* $J(n) = \begin{cases} 1 & \textit{if } f(n) < u \\ 0 & \textit{otherwise} \end{cases}$,

where u *is a continuity point of* δ *, then,* I *and* J *are associated.*

<u>Remark</u>. This theorem is another formulation of the result given in [41] (P.16). $\delta(u)$, when it exists, is "the distribution function of f ". In fact, the relation with the "distribution of F in u ", i.e. $m\{t|F(t) < u\}$, is clear only when u is a continuity point of δ . Most of the difficulties arise from the following phenomenon:

We consider the case of $F = 0$; to F , we associate $f(n) = \frac{(-1)^n}{n}$; then, $M(|f|) = 0$ and

$$\delta(u) = \lim_{x\to+\infty} \frac{1}{x} \sum_{\substack{n\leq x \\ f(n)<u}} 1 = \begin{cases} 0 & \text{if } u < 0 \\ \frac{1}{2} & \text{if } u = 0 \\ 1 & \text{if } u > 0 . \end{cases}$$

Clearly the only corresponding "distribution function" in the probability sense will be $\sigma(u)$ defined by

$$\sigma(u) = \begin{cases} 0 & \text{if } u \leq 0 \\ 1 & \text{if } u > 0 \end{cases} ,$$

and this is the distribution function of F . Since $\delta(0) = \frac{1}{2}$, δ is not a distribution function in the ordinary sense. The only possibility to avoid this critical situation is to restrict to continuity points, for, by such a choice, the continuity of δ will imply the continuity of the distribution function of F since the two are approximated by the same sequence of functions, in $B^{\alpha}L$-P and in $^{\alpha}(G,dm)$ respectively.

3-A-3. Some consequences of the properties of the probability spaces (continued): a special case.

As it has been explained in III-A-1-b, if we deal only with functions depending on divisors of integers, like additive or multiplicative arithmetical functions, the basic facts are the following:

$$\text{Let } E = \prod_p ([1,p,p^2,\cdots[\cup\{\infty\}_p) .$$

E can be identified to $(G\setminus\{0\})/G^*$, and the measure $d\mu$ induced by dm on E is

$$d\mu = \bigotimes_p d\mu_p ,$$

where $d\mu_p$ is defined by $\mu_p(\{p^k\}) = (1 - \frac{1}{p}) \cdot \frac{1}{p^k}$.

The correspondence between $L^\lambda(E,d\mu)$ and B^λ-L-P shows that the Fourier series of an element F in $L^\lambda(E,d\mu)$ can be written by

$$F(t) \sim \sum_{q=1}^{+\infty} \hat{F}_q \, C_q(t) ,$$

where $C_q(t)$ is the extension of the qth-Ramanujan sums on E, which is uniquely defined.

The set of real trigonometric polynomials defined on G and invariant under G^* can be identified to the set Γ :

$$\Gamma = \{\gamma: \mathbb{N}^* \to \mathbb{R} \mid \gamma = \sum_{\substack{d<+\infty \\ d\in\mathbb{N}^*}} \lambda_d I_d , \quad \lambda_d \in \mathbb{R}\} ,$$

$$\text{where } I_d(n) = \begin{cases} 1 & \text{if } d|n \\ 0 & \text{otherwise} \end{cases} .$$

As it was remarked by J. Knopfmacher [21], it is possible to extend such a construction to rather general semi-groups. The present author would like to point out that this kind of construction can be done in any compact set which is a projective limit of finite sets.

3-B. Applications.

We shall restrict ourself to only three applications. The first one is related to the classical topic of multiplicative arithmetical functions; the second one, to the extension of an arithmetical multiplicative function in an algebraic number field; the third one will give the solution of some problems of W. Schwarz concerning the foundations of the theory of almost-periodical sequences. Our purpose is to show that it is possibly not only to get new results, but also to understand their meaning, when we refer to the abstract measure theoretic viewpoint. These applications of measure theory to number theory will provide a better understanding of some properties of arithmetical functions.

3-B-1. Multiplicative arithmetical functions.

A multiplicative arithmetical function f is a function $f : \mathbb{N}^* \to \mathbb{C}$ such that $f(mn) = f(m) \cdot f(n)$ if $(m,n) = 1$.

Let f be a multiplicative arithmetical function. Denote by ζ the Riemann Zeta function. We consider the following conditions:

c_1: $\lim\limits_{\sigma \to 1^+} \zeta(\sigma)^{-1} \cdot \left(\sum\limits_{n=1}^{+\infty} \frac{f(n)}{n^\sigma} \right)$ exists and is not 0 .

c_2: For any $\varepsilon \geqq 0$, there exists $\eta \geqq 0$ such that if g belongs to Γ and satisfies

$$\lim_{\sigma \to 1^+} \zeta(\sigma)^{-1} \cdot \sum_{n=1}^{+\infty} \frac{|g(n)|}{n^\sigma} < \eta .$$

Then we have:

$$\lim_{\sigma \to 1^+} \zeta(\sigma)^{-1} \cdot \sum_{n=1}^{+\infty} \frac{|f(n)g(n)|}{n^\sigma} < \varepsilon .$$

We have the following result:

Theorem. *A multiplicative arithmetical function* f *satisfies* c_1 *and* c_2 *if and only if*

$\prod_{p \leq y} f(p^{V_p(t)})$ *tends to a limit* $F(t)$ *almost-everywhere and in* $L^1(E, d\mu)$, *and* $\int_E F \, d\mu$ *is not* 0 .

Remark. It is interesting to compare with the following result:

Theorem. *Let* g *be a multiplicative arithmetical function. Then,* $\lim_{y \to +\infty} \prod_{p \leq y} g(p^{V_p(t)})$ *exists in* $L^1(G, d\mu)$ *and* dm-*almost everywhere if and only if there exists a Dirichlet character* χ *such that* χg *satisfies the above theorem.*

For the proof of these theorems, we refer to [41].

It is not difficult to give a more general result in the case where c_1 is replaced by c_1' :

$$c_1' : \limsup_{\sigma \to 1} \left| \zeta(\sigma)^{-1} \times \sum_{n=1}^{+\infty} \frac{f(n)}{n^\sigma} \right| > 0 .$$

In this case, we get:

$$\lim_{y \to +\infty} \prod_{p \leq y} \frac{f(p^{V_p(t)})}{(1 - \frac{1}{p}) \sum_{k=o}^{+\infty} \frac{f(p^k)}{p^k}} \text{ exists almost everywhere and in}$$

$L^1(E, d\mu)$. In such a case, an easy consequence is:

$$\lim_{x \to +\infty} \left\{ \frac{1}{x} \sum_{n \leq x} f(n) \right\} \times \frac{1}{\prod_{p \leq x} (1 - \frac{1}{p}) \sum \frac{f(p^k)}{p^k}} = 1 .$$

N.B. This result can be proved without the analytic method of Halasz. For a proof in the special case $|f| \leq 1$, see [42].

We shall stop here to deal with the first topic, and go to the second one.

3-B-2. Extension of arithmetical multiplicative functions into an algebraic number field.

Let K be an algebraic number field and $I(K)$ be the set of the integral ideals of K. A function $H : I(K) \to \mathbb{C}$ is multiplicative if $H(AB) = H(A) \cdot H(B)$ if $(A,B) = 1$.

In the same way as in the case of integers, it is possible to define the topological space E_K by $E_K = \prod_P E_{P,K}$, where P belongs to the set of the prime ideals of E, and $E_{P,K}$ is the compactification of $[1, P, P^2, \cdots[$; on E_K, we define a measure $d\mu_K$ by:

$$d\mu_K = \bigotimes_P d\mu_{P,K} ,$$

where $d\mu_{P,K}(P^\alpha) = (1 - \frac{1}{N(P)}) \times \frac{1}{N(P)^\alpha}$, $\alpha \geq 0$, and $N(\cdot)$ = norm of $(\cdot)$.

We shall denote by $M(\mathbb{N}^*)$ (resp. $M(K)$) the set of multiplicative functions on $\mathbb{N}^*$ (resp. on $I(K)$).

Now, given h in $M(\mathbb{N}^*)$, we say that h is commutatively integrable if for any reordering π of P, the set of prime numbers, we have

$$\lim_{y \to +\infty} \prod_{\substack{p_j \in \pi(P) \\ j \leq y}} h(p_j^{V_{p_j}(t)}) \text{ exists in } L^1(E, d\mu) .$$

Clearly, such a definition can still be applied to the case of elements of $M(K)$, with obvious modifications of notations. We shall denote by $C.L(E,d\mu)$ (resp. $C.L(E_K, d\mu_K)$) the set of commutatively integrable elements of $M(\mathbb{N}^*)$ (resp. $M(K)$).

The result is the following:

Theorem. *Given any algebraic number field* K *on* $\mathbb{Q}$, *there exists a natural extension* e_K *of* $M(\mathbb{N}^*)$ *to* $M(K)$ *such that:*

- e_K *preserves the Dirichlet convolution* $*$, *i.e.*
  $e_K(f * g) = e_K(f) * e_K(g)$,
- $e_K(M(\mathbb{N}^*)) \subset M(K)$

and

- $e_K(C.L(E,d\mu)) \subset (C.L(E_K,d\mu_K))$.

N.B. 1. The description of this extension has been given in [45], with some properties. The proof that $e_K(CL(E,d\mu)) \subset CL(E_K,d\mu_K)$ is rather straightforward, since the essential point here is the fact that h belongs to $CL(E,d\mu)$ if and only if

$$\sum_p \frac{|1-f(p)|}{p} < +\infty \ , \quad \sum_{k\geq 2} \sum_p \frac{|f(p^k)|}{p^k} < +\infty \ .$$

2. If we do not ask for h in $CL(E,d\mu)$, but only for h in $L(E,d\mu)$, the theorem is not true; it is not difficult to investigate this case and give results in this special case. Anyway, it is not sure that $C.L(E,d\mu)$ is the maximal subspace S of $L(E,d\mu)$ such that $e_K(S) \subset S_K$, where S_K is the corresponding subspace in $L(E_k,d\mu_k)$.

3. The theorem solves the "going-up" from $\mathbb{N}^*$ to $I(K)$. The "going-down" problem can be solved also, but it is rather tricky and we shall not treat it here.

3-B-3. <u>On some problems of W. Schwarz</u>.

a) Let $f : \mathbb{N}^* \to \mathbb{C}$, of B^1-L-P with Fourier Series

$$f(n) \sim \sum_{q=1}^{+\infty} a_q(f)\, c_q(n) \ .$$

If f is an additive or multiplicative function, this series converges to $f(n)$ for any n in $\mathbb{N}^*$. The same result holds if f is U.L.P.[43]. In 3-2-b, N.B. ii), it was proved that, for N_k satisfying some hypothesis,

$$f_k(t) = \sum_{q|N_k} a_q(f)\, c_q(t)$$

converges $d\mu$.a.e. and in $L(E,d\mu)$. It may be remarked that, if $f \in$U.L.P, then f_k converges uniformly to f . Now, if f is additive or multiplicative, then, $f_k(n)$ converges to $f(n)$ for any n in $\mathbb{N}^*$; the proof is really immediate. The reason for which such a phenomenon occurs is that these three kinds of functions are completely determined by their values on $\mathbb{N}^*$, and any modification of one of these values modifies the whole function. Now, it is clear that there is no hope to find a general summability method for the Fourier Series of an element of $B^1.L.P$ to this element, since such a method will give one representative of the class of f, and there is no a priori reason for which this representative will be f itself. Of course, the summation by the f_k is more efficient than the ordinary one, since it saves the uniformity, for instance. But it seems that the "problem of giving a "simple" summability method summing the series (*) for "many" f in B.L.P." (Pb. 8.1, p.49 in [43]) will admit a solution if, first of all, the choice of the representative of the class of f is allowed. In this case, the "summability method" is immediate; we refer to 3-2-b, N.B.; first, select a sequence N_k with non-decreasing $V_p(N_k)$ satisfying $\lim_{k\to\infty} V_p(N_k) = +\infty$ for any p in $\mathcal{P}$; next, define f_k by :

$$f_{k+1}(n) = \sum_{q|N_k} a_q(f)\, c_q(n) ;$$

if $f(n)$ is defined by

$$f(n) = f_{k+1}(n) \quad \text{if} \quad N_k < n \leqq N_{k+1} ,$$

a natural summability method arises immediately: given m in $\mathbb{N}^*$, let k be an integer with $N_k < m \leq N_{k+1}$, then

$$\sum_{q|N_k} a_q(f)\, c_q(m) = f(m) ,$$

by construction.

So, this problem is solved, and we go to the next one; f_k and f have the same meaning as just above.

b) It is easy to show that, for given N_k such that $V_p(N_k)$ is non-decreasing and tends to $+\infty$, as $k \to +\infty$ for every p in $\mathcal{P}$, the sequence f_k converges to f uniformly if f is U.L.P. From this fact, it is immediate that since $f_k(n) = f_k((n,N_k))$, where

(n,N_k) is the G.C.D. of n and N_k, we get: $f((n,N_k))$ converges uniformly to f if f is U.L.P.

The extension to the case of B.L.P. sequences is not possible; this means:

There exists B.L.P. sequences f such that, for some sequences N_k satisfying the above conditions, we do not have:

$$\lim_{k\to+\infty} M|f(n) - f((n,N_k))| = 0 .$$

We give an example: Define f by:

$$f(n) = \begin{cases} 2^r h(r) & \text{if } n = 2^r \\ 0 & \text{otherwise} \end{cases},$$

with $h(r) \to 0$, as $r \to +\infty$. Then we have:

$$M(f) = \lim_{x\to+\infty} \frac{1}{x} \sum_{r=0}^{\frac{\log x}{\log 2}} 2^r h(r) = 0 .$$

Now, if $\sum_r h(r) = +\infty$, $h(r) > 0$ and denoting by $g(k)$ the expression $\sum_{r\leqq k} h(r)$, we define N_k by $N_k = 2^k \times \prod_{2<p\leqq g(k)} p^k$. We have:

$$\frac{1}{x} \sum_{n=1}^{x} f((n,N_k)) \to \frac{1}{N_k} \sum_{n=1}^{N_k} f((n,N_k)) , \quad x \to +\infty .$$

But

$$\sum_{n=1}^{N_k} f((n,N^k)) \geqq \sum_{r=0}^{k} f(2^r) \times \sum_{\substack{(m,N_k)=1 \\ m\leqq \frac{N_k}{2^r}}} 1$$

$$\geqq \sum_{r=0}^{k} f(2^r) \times \frac{N_k}{2^r} \times \prod_{p|N_k} (1 - \frac{1}{p})$$

$$\geqq (\sum_{r=0}^{k} h(r)) \times \frac{e^\gamma}{\log g(k)} \times N_k(1+o(1)) , \quad k \to +\infty$$

Hence we get:

$$\frac{1}{N^k} \sum_{n=1}^{N_k} f((n,N_k)) \to +\infty \ , \ k \to +\infty \ .$$

This shows that the extension of the result in the case of U.L.P. sequences associated to continuous functions on E to B.L.P. sequences is not possible. (This question was asked about in [43] (see problem 8.2, p.49)).

c) The characterization of elements of B.L.P. originally given by Novoselov and recalled at the beginning of this paper, together with the construction explained above, permits to give a characterization of sequences associated to elements of $L(E,d\mu)$. Anyway, the "structural" or "number theoretical" properties of these sequences are clearly related to the structure of $(G\setminus\{0\})/G^*$, and any characterization of these sequences will be direct a translation of meaure-theoretical properties. (This is a comment on the problem 8.3, p.49 in [43].)

d) If we look at the spectrum of a Besicovitch-almost-periodical sequence, it is a subset of $\mathbb{R}/\mathbb{Z}$. Now, remark that $\mathbb{R}/\mathbb{Z}$ is isomorphic to $(\mathbb{Q}\cap[0,1[) \times (\mathbb{R}/\mathbb{Q})$. So, the dual group of $\mathbb{R}/\mathbb{Z}$ is isomorphic to the product $G \times H$, where G = dual of $(\mathbb{Q}\cap[0,1[) = \prod_p \mathbb{Z}_p$, H = dual of $\mathbb{R}/\mathbb{Q}$. A uniformly almost periodical sequence is the restriction to $\mathbb{N}^*$ of a continuous function on $G \times H$ with complex value. Of course, if f is U.L.P., it is the restriction to $\mathbb{N}^*$ of a continuous function on G . (This solves problem 8.4, p.49 in [43].)

Remark. The relation between Gelfand theory and almost periodicity is nicely explained in [44]. In the present case, the viewpoint of A. Weil [33] is sufficient for solving the problem.

e) If we have a Besicovitch-almost-periodical sequence $f(n)$, it is possible to analyze its components very easily. In fact, we select a sequence N_k with the properties: $N_k | N_{k+1}$, and for any p , $\lim_{k\to+\infty} V_p(N_k) = +\infty$.

Consider:

$$\lim_{x\to+\infty} \frac{1}{x} f(n+N_k m) = f_k(n) .$$

$f_k(n)$ is periodical and a Cauchy sequence in B.L.P. So, to f_k, we associate a limit f' . f' is a representative of the class of the B.L.P. component of f , i.e.: $(f-f')$ is Besicovitch-almost-periodical with irrational spectrum.

Now, if we take

$$\tilde{f}_k(n) = \frac{1}{\varphi(N_k)} \sum_{\substack{m=1 \\ (m,N_k)=1}}^{N_k} f_k(nm) ,$$

where φ is the Euler function, then $\tilde{f}_k(n)$ is a component of $f_k(n)$ invariant under G^* . Since it is a Cauchy sequence in B.L.P., the limit f'' exists and is a representative of the component of f' invariant under G^* . So, a Besicovitch-almost-periodical sequence f can be written:

$$f = (f-f') + (f'-f'') + f'' ,$$

where $(f-f')$ has an irrational spectrum,

$(f'-f'')$ has no invariant component by G^* , with a rational spectrum, and is B.L.P.

f'' is invariant almost-everywhere by G^* , and B.L.P.,

and the constructions have been effectively described. (This provides a solution to problem 8.5, p.49 in [43].)

Remark. Problems 8.6 and 8.7 of [43] are not related to the same topic as here. But problem 8.8 does. We gave its solution in [46], and shall not repeat it here.

Conclusion. Starting from the underlying structures associated to the truncation method, we have been able to give a new viewpoint on the theory of arithmetical functions, and to solve some related problems. It is possible to deal with more general structures, for instance, with the ring of square 2×2 matrices with integral entries, which will be discussed elsewhere.

REFERENCES

[1] Cesaro, E.: Probabilités de certains faits arithmétiques, Mathesis, 11, 150-151 (1884).

[2] Cesaro, E.: Eventualité de la division arithmétique, Ann. Mat. pura e appl., (2) 13, 295-313 (1889).

[3] Steinhaus, H.: Les probabilités dénombrables et leurs rapports à la théorie de la mesure, Fund. Math., 4, 286-310 (1923).

[4] Kolmogorov, Grundbegriffe der Wahrscheinlichkeitsrechnung, Springer, Berlin, 1933.

[5] Erdös, P. and Kac, M.: The Gaussian law of errors in the theory of additive number-theoretic functions, Amer. J. Math., 62, 738-742 (1940).

[6] Kubilius, J., Probabilistic Methods in the Theory of Numbers, A.M.S. Translations n°11, Providence, 1964.

[7] Elliott, P.T.T.A., Probabilistic Number Theory, I, II, Springer, Berlin, 1979.

[8] Wintner, A., Eratosthenian average, Baltimore, Md., 1943.

[9] Delsarte J.: Essai sur l'application de la théorie des fonctions presque-périodiques à l'arithmétique, Annales Sci. Ecole Norm. Sup., (3) 62, 185-204 (1945).

[10] Erdös, P. and Wintner, A.: Additive arithmetical functions and statistical independence, Amer. J. Math., 61, 713-721 (1939).

[11] Erdös, P. and Wintner, A.: Additive functions and almost-periodicity (B^2), Amer. J. Math., 62, 635-645 (1940).

[12] Kac, M., Van Kampen, E.R. and Wintner, A.: Ramanujan sums and almost-periodic functions, Amer. J. Math., 62, 107-114 (1940).

[13] Van Kampen, E.R.: On uniformly almost-periodic multiplicative and additive functions, Amer. J. Math., 62, 627-634 (1940).

[14] Novoselov, E.V.: Integration on a bicompact ring and its applications to number theory (Russian), Izv. Vyss, Ucebn. Zaved. Matematika, 22, n°3, 66-79 (1961).

[15] Novoselov, E.V.: Foundations of classical analysis and of the theory of analytic functions in a polyadic region (Russian), Isv. Vyss. Ucebn. Zaved. Matematika, 36, n°5, 71-88 (1963).

[16] Novoselov, E.V.: A new method in probabilistic number theory, Izv. Akad. Nauk SSSR ser. Mat., 28, 307-364 (1964).

[17] Mautner, F.I.: On congruence characters, Monat. für Math., 57, (4), 307-316 (1953).

[18] Følner, E.: On the dual spaces of the Besicovitch almost periodic spaces, Danske Vid. Selsk. Mat.-Fys. Medd., 29, n°1, 1-27 (1954).

[19] Cohen, E.: Almost even-functions of finite abelian groups, Acta. Arith., 7, 311-323 (1961/62).

[20] Schwarz, W. and Spilker, J.: Eine Anwendung des Approximationssatzes von Weierstrass-Stone auf Ramanujan-Summen, Nieuw Arch. Wiskunde, (3) 19, 198-209 (1971).

[21] Knopfmacher, J.: Fourier analysis of arithmetical functions, Ann. Mat. pura Appl., (4) 109, 177-201 (1976).

[22] Schwarz, W. and Spilker, J.: Mean values and Ramanujan expansions of almost-even arithmetical functions, Topics in Number Theory, Colloq. Math. Soc. Jonas Bolyai, 13, 315-357 (1976).

[23] Kryzius, Z.: Additive arithmetic functions on semigroups and the preservation of weak convergence of measures, Liet. Mat. Rinkinys, 25, n°1, 72-83 (1985).

[24] Kryzius, Z.: Almost-even arithmetic functions on semigroups, Liet. Mat. Rinkinys, 25, n°2, 90-101 (1985).

[25] Kryzius, Z.: Limit periodic arithmetic functions, Lith. Math. J., 243-250 (1986).

[26] Van Dantzig, D.: Uber topologisch homogene Kontinua, Fund. Math., 15, 102-125 (1930).

[27] Van Dantzig, D.: Zur topologischen Algebra, I, Komplettierungstheorie, Math. Ann., 107, 587-626 (1935).

[28] Van Dantzig, D.: Zur topologischen Algebra, II, Abstrakte b_ν-adische Ringe, Compositio Math., 2, 201-223 (1935).

[29] Van Dantzig, D.: Nombres universels ou $\nu!$-adiques avec une introduction sur l'algèbre topologique, Ann. Sci. Ecole Norm. Sup., (3) 53, 275-307 (1936).

[30] Van Dantzig, D.: Zur topologischen Algebra, III, Brouwersche und Cantorsche Gruppen, Compositio Math., 3, 408-426 (1936).

[31] Mauclaire, J.-L.: Suites limite-périodiques et théorie des nombers, VI, Proc. Japan Acad., 57 (A), 223-225 (1981).

[32] Abe, M.: Über Automorphismen der lokal-kompakten abelschen Gruppen, Proc. Imp. Acad. Tokyo, 16, 59-62 (1940).

[33] Weil, A.: L'intégration dans les groupes topologiques et ces applications, Actualités Sci. et Ind., 869, Paris, Hermann et Cie (1941).

[34] Buck, R.C.: The measure-theoretic approch to density, Amer. J. Math., 68, 560-580 (1946).

[35] Bohr, H. and Følner E.: On some types of functional spaces. A contribution to the theoty of almost-periodic functions, Acta Math., 76, 31-155 (1944).

[36] Hartman, P. and Wintner A.: On the almost periodicity of additive number-theoretical functions, Amer. J. Math., 62, 753-758 (1940).

[37] Kac, M.: Almost periodicity and the representation of integers as sums of squares, Amer J. Math., 62, 122-126 (1940).

[38] Mauclaire, J.-L., Sur les travaux de Novoselov, Groupe de Théorie Analytique des Nombres, I.H.P., Paris, 1985-86.

[39] Marcinkiewicz J.: Une remarque sur les espaces de M. Besikowitch, C.R. Acad. Sci., Paris, 208, 157-159 (1939).

[40] Knopfmacher, J.: Abstract analytic Number Theory, North-Holland, Amsterdam-Oxford, 1975.

[41] Mauclaire, J.-L., Intégration et Théorie des Nombres, Hermann, Paris, 1986.

[42] Mauclaire, J.-L.: Probability spaces in number theory: an example, Proceedings of the Vth Soviet-Japan symposium on Probability theory, Kyoto, Springer (1986), to appear.

[43] Schwarz., W.: Remarks on the theorem of Elliott and Daboussi and applications, Extended version of two survey lectures given at the Warszawa International Banach Center on 6-9-1982, 89 (1982), (Preprint).

[44] Loomis, L.H., Abstract Harmonic Analysis, Van Nostrand, Princeton, 1953. (See Ch. VII, §41).

[45] Mauclaire, J.-L.: On the extention of a multiplicative arithmetical function in an algebraic number field, Math. Japonicae, 21, 337-342 (1976).

[46] Mauclaire, J.-L.: Deux prolèmes sur les suites limite-périodiques, Proc. Japan Acad., 63, A, 90-93 (1987).

Proc. Prospects
of Math. Sci.
World Sci. Pub.
127-140, 1988

SOME APPLICATIONS OF THE THEORY OF AUTOMATA

MICHEL MENDES FRANCE

Department of Mathematics
University Bordeaux I
33405 Talence cēdex
France

1. FOUR PROBLEMS

Problem 1 (complexity theory). Is the binary expansion of $\sqrt{2}$ random?

$$\sqrt{2} = 1.0110101000001001111001100110011111110011101111001100\cdots .$$

The notion of randomness is a problem in itself. The naive question could be to ask whether the digits 0 and 1 occur with frequency 1/2, whether each couple 00,01,10,11 occurs with frequency 1/4, etc.... This question seems extremely difficult. We do not even know whether the couple 00, say, appears infinitely often [10]! We shall nevertheless give a partial answer to our problem in terms of complexity (see also [16]).

Problem 2 (Number Theory). Folding a sheet of paper repeatedly over and over again (Shunji Ito insists that I should read "Gama-no-Abura-uri") generates an infinite sequence of folds V and Λ.

VVΛVVΛΛVVVΛΛVΛΛ .... .

Put V=1 and Λ=0. Is the real number (paperfolding number)

$$.110110011100100 \cdots$$

transcendental? [13]

Problem 3 (Number Theory). Find the continued fraction expansion of the Fredholm number:

$$\varphi = \sum_{n=0}^{\infty} g^{-2^n} ,$$

where $g \geqq 2$ is an integer [20].

Problem 4 (One dimension physics). Describe the ground state of an inhomogeneous one dimensional Ising chain [2],[3].

We intend to show that all four problems can be studied by using the theory of automata. For further applications we refer to a recent paper of J.P. Allouche [1] and to an older article of Dekking, Mendès France, Van der Poorten [14].

2. WHAT IS A k-AUTOMATON?

Let us first describe a 2-automaton. It consists of a finite number of states A,B,C,···. One of these states, say A, is the initial state.

Each state is joined to two states (not necessarily distinct) by arrows named 0 and 1. Finally each state is mapped on a new alphabet a,b,c,··· .

For example,

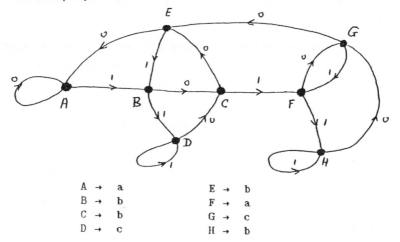

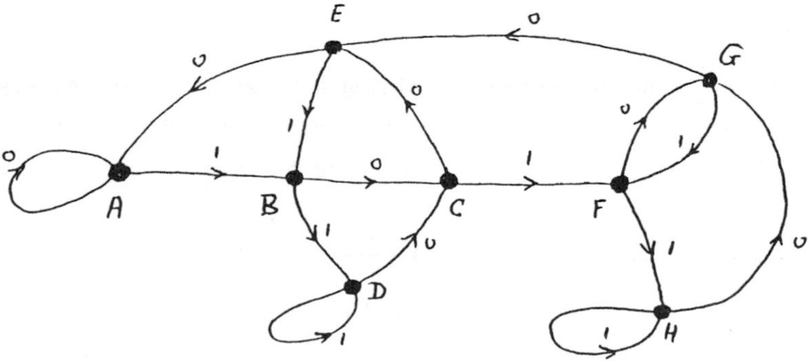

A → a		E → b	
B → b		F → a	
C → b		G → c	
D → c		H → b	

This particular automaton $\mathcal{A}$ maps "nineteen" = 10011 in binary expansion into the state D, starting from A and following the instructions 1,0,0,1,1; we read off c. In the same fashion, every integer $n \geq 0$ is sent into a symbol $a_n \in \{a,b,c\}^{\mathbb{N}}$. We say that the sequence $(a_n) \in \{a,b,c\}^{\mathbb{N}}$ is generated by the automaton $\mathcal{A}$. Here are the first terms of (a_n).

n	0	1	10	11	100	101	110	111	1000	1001	1010	1011	1100	1101	···
st	A	B	C	D	E	F	C	D	A	B	G	H	E	F	···
a_n	a	b	b	c	b	a	b	c	a	b	c	b	b	a	···

Let $k \geq 2$ be a given integer. A k-automaton is defined as above except that the inputs are integers expressed to the base k and that the k arrows leaving from each state are named 0, 1, 2, ···, k-1.

A sequence generated by a k-automaton is called "k-automatic", or simply "automatic" if no confusion is possible.

For a given infinite sequence on a finite alphabet (with 2 letters at least), one can ask whether it is automatic. This question is pertinent since the family of automatic sequences is countable whereas the family of all sequences on the same alphabet is uncountable.

An ultimately periodic sequence is automatic but there do exist automatic sequences which are not ultimately periodic. The family of automatic sequences is stable under shift, under modification of finitely many terms, etc. ··· .

We now give some typical examples of 2-automatic sequences.

Example 1: Thue-Morse sequence.

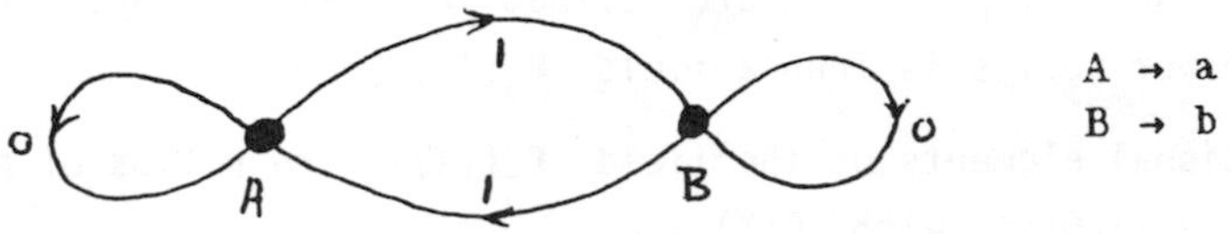

This sequence coincides with $((-1)^{s(n)})$, $n \in \mathbb{N}$, where $s(n)$ is the number of 1's in the binary expansion of n.

Example 2: Fredholm sequence.

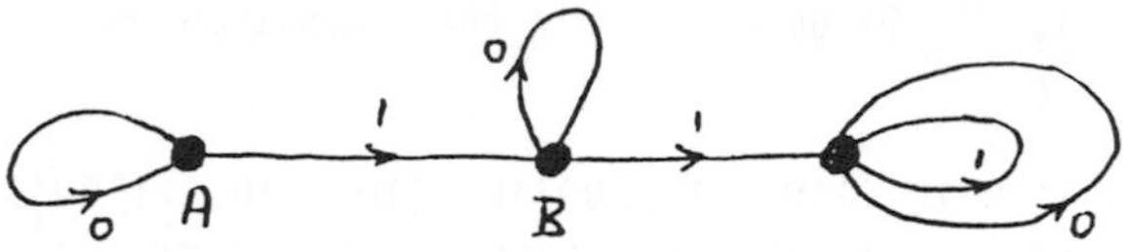

A ↦ a
B ↦ b
C ↦ a

This sequence picks out the powers of 2:

abbabaaabaaaaaaaba ••• .

Example 3: Paperfolding sequence.

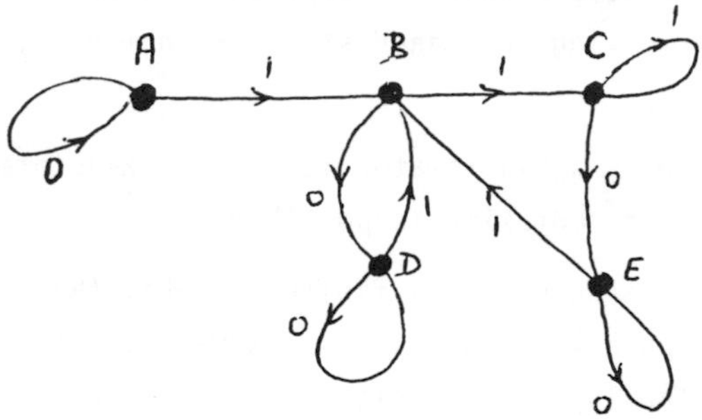

A ↦ ø
B ↦ ∨
C ↦ ∧
D ↦ ∨
E ↦ ∧

3. A CHARACTERIZATION OF AUTOMATIC SEQUENCES

Let p be a prime and $q=p^{\nu}$ ($\nu \geq 1$ integer). $\mathbb{F}_q$ denotes the Galois field with q elements. A formal power series $f=f(X)$ on $\mathbb{F}_q$ is of the form:

$$f = f(X) = \sum_{n=n_0}^{\infty} f_n X^n, \quad f_n \in \mathbb{F}_q, \ n_0 \in \mathbb{Z} .$$

Addition and multiplication are defined in a natural way. The set of formal power series is then a field $\mathbb{F}_q((X))$.

Rational elements of the field $\mathbb{F}_q((X))$ are ratios of polynomials $P(X)/Q(X)$, with $Q(X) \neq 0$.

It is quite obvious that

$$f = \sum_{n=n_o}^{\infty} f_n X^n$$

is rational if and only if the sequence of coefficients (f_n) is ultimately periodic. The family of rational elements is denoted $\mathbb{F}_q(X)$.

An element $f \in \mathbb{F}_q((X)$ is algebraic (over $\mathbb{F}_q(X)$) if there exist rational elements $a_{d-1}, a_{d-2}, \cdots, a_o$ such that

$$f^d + a_{d-1} f^{d-1} + \cdots + a_1 f + a_o = 0.$$

If we wish to specify the ground field, we say that f is q-algebraic. An element which is not q-algebraic is said to be q-transcendental.

Theorem 1 (Christol, Kamae, Mendès France, Rauzy [12])
Let $(f_n)_0^{\infty}$ *be an infinite sequence on the alphabet* $\mathbb{F}_q$. *The sequence is* q-*automatic if and only if the series*

$$f = \sum_{n=o}^{\infty} f_n X^n \quad \textit{is q-algebraic.}$$

For the proof we refer to [12].

Example 1. The Thue-Morse sequence on the alphabet $\{0,1\}$ corresponds to the Thue-Morse series f in $\mathbb{F}_2((X))$ which satisfies

$$(1+X)^3 f^2 + (1+X)^2 f + X = 0.$$

Example 2. The Fredholm sequence $a=0$, $b=1$ corresponds to the series

$$f = \sum_{n=o}^{\infty} X^{2^n},$$

which satisfies (in $\mathbb{F}_2((X))$

$$f^2 + f + X = 0.$$

Example 3. The paperfolding series f is a solution of

$$(1+X)^4 f^2 + (1+X)^4 f + X = 0 ,$$

in $\mathbb{F}_2((X))$.

4. TRANSCENDENTAL ELEMENTS

Let $p \neq p'$ be two primes and suppose that $q = p^{\nu} < q' = p'^{\mu}$ with $\nu \geqq 1$, $\mu \geqq 1$. The set $\mathbf{F}_{q'}$ is larger than the set $\mathbf{F}_q$. Let i be an injection $\mathbf{F}_q \overset{i}{\to} \mathbf{F}_{q'}$.

The injection can be lifted to an injection

$$\mathbf{F}_q((X)) \to \mathbf{F}_{q'}((X)),$$

so that an element of $\mathbf{F}_q((X))$ can be considered as an element in $F_{q'}((X))$. The injection is a set injection. It is never a homomorphism!

Theorem 2 (change of base) [12]. *If* $f \in \mathbf{F}_q((X))$ *is irrational and* q-*algebraic then it is* q'-*transcendental.*

This result is a mere translation of a theorem of Cobham's [13] which asserts that if a sequence is q-automatic and q'-automatic, then it must be ultimately periodic.

The change of base destroys the algebraic nature of an irrational element. This seems to be a general principle which led J. Loxton and A. Van der Poorten to establish the following difficult theorem [17].

Theorem 3 (Loxton, Van der Poorten). *Let* $\mathbf{F}_q \subset Z$ *be an injection. If* $f = \sum_{n=0}^{\infty} f_n X^n \in \mathbf{F}_q((X))$ *is an irrational* q-*algebraic element, and is "subject to some technical requirement", then the number:*

$$\sum_{n=0}^{\infty} f_n \alpha^n \in \mathbf{C}$$

is transcendental for all nonzero complex algebraic number α *provided* $|\alpha| < 1$.

The "technical requirement" is described in detail in [17]. The authors expect to be able to prove that the condition is actually unnecessary.

This last result has two strinking consequences.

Corollary 1. *The paperfolding number is transcendental.*

Indeed, Example 3 shows that the paperfolding series is 2-quadratic. We have thus solved Problem 2. The technical requirement happens to be fulfilled in this Example. The answer to Problem 1 is given by the following statement.

Corollary 2. *Let* ξ *be a real irrational algebraic number, the expansion to the base* g *of which is*

$$\xi = \sum_{n=o}^{\infty} \frac{\xi_n}{g^n}, \quad \xi_n \in \{0,1,\cdots,g-1\}.$$

The sequence (ξ_n) *is not automatic hence the sequence is random!*

Indeed, the formal power series:

$$\xi\ (X) = \sum_{n=o}^{\infty} \xi_n\ X^n \in \mathbb{F}_q\ ((X)), \quad \text{for} \quad q \geqq g$$

is q-transcendental. If it were q-algebraic, the real number would be transcendental!

By theorem 1, this implies that (ξ_n) is not q-automatic. What we state here, the sequence is random, is just a definition! Yet, our result does say something, namely that the sequence of digits (ξ_n) cannot be generated by a simple machine.

The proof is unfortunately incomplete since it depends on the "technical requirement". Strictly speaking Corollary 2 is still a conjecture.

5. A CONTINUED FRACTION

Let $a/b \in (0,1)$ be a rational number which we expand into continued fraction:

$$\frac{a}{b} = \cfrac{1}{\alpha_1 + \cfrac{1}{\alpha_2 + \cdots + \cfrac{1}{\alpha_s}}} = [0,\alpha_1,\alpha_2,\cdots,\alpha_s] .$$

The partial quotients α_i's are positive integers. Without loss of generality we may assume s to be even with $\alpha_s \geq 1$. Observing the perturbed symmetry pattern:

$$\frac{a}{b} + \frac{1}{b^2} = [0,\alpha_1,\cdots,\alpha_{s-1},\alpha_s+1,\alpha_s-1,\alpha_{s-1},\cdots,\alpha_1].$$

M. Kmosek and J. Shallit [21] independently discovered the structure of the continued fraction expansion of the Fredholm number

$$\mathcal{F} = \sum_{n=o}^{\infty} \frac{1}{g^{2^n}} \quad (g \geq 2 \text{ integer}).$$

As a matter of fact

$$\frac{1}{g} + \frac{1}{g^2} = [0,g-1,g+1].$$

Apply the perturbed symmetry principle to obtain

$$\frac{1}{g} + \frac{1}{g^2} + \frac{1}{g^{2^2}} = [0,g-1,g+2,g,g-1],$$

then

$$\sum_{n=o}^{3} \frac{1}{g^{2^n}} = [0,g-1,g+2,g,g,g-2,g,g+2,g-1],$$

and after infinitely many applications of the principle:

$$\mathcal{F} = [0,g-1,g+2,g,g,g-2,g,g+2,g,g-2,g+2,g,g-2,\cdots].$$

It is an amusing exercise to show that the sequence of partial quotients is 2-automatic [11]. In fact deleting the first partial quotient $g-1$, the sequence coincides with the sequence generated by the 8 state automaton $\mathcal{A}$ described in Section 2 by choosing $a=g+2$, $b=g$, $c=g-2$.

Kmosek and Shallit created some excitement since they gave an example of a real number the decimal expansion by taking $g=10$ and the continued fraction expansion of which are "explicitly" known. Only a few other examples are known.

By making use of this fact, I was able to solve a special case of an inhomogeneous Ising chain [19] (which will not be discussed here). However our next Section will be devoted to the Ising chain though in a different context.

Let us make a last remark on the Fredholm number $\mathcal{G}$. Our Example 3 in Section 3 shows that the Fredholm series.

$$f = \sum_{n=o}^{\infty} X^{2^n} \in \mathbb{F}_2((X))$$

is irrational and 2-algebraic. Hence $\mathcal{G}$ is a real transcendental number. Actually this has been known since K. Mahler [18]. An old problem which seems extremely difficult is to decide whether all real numbers whose partial quotients are bounded are either quadratic (ultimately periodic partial quotients) or transcendental. The answer seems out of reach. An easier problem could be to show that if the sequence of partial quotients is automatic and not ultimately periodic, then the number is transcendental. See A. Baker [7] for related results.

Some very interesting and recent work has been developed around the theory of continued fractions in the field $\mathbb{F}_q((X^{-1}))$.

Baum and Sweet [8], [9] on the one hand and Mills and Robbins [20] on the other hand produced q-algebraic elements of degree $d \geqq 3$ which have bounded partial quotients. The structure of the sequence of partial quotients seems slightly more complex than that of automatic sequences. See however the recent work of Allouche, Betrema and Shallit [5].

6. THE ISING CHAIN

The Ising chain is a one dimension model of magnetic substance. We are given N sites on which are defined spins $\sigma_n = \pm 1$ with $1 \leqq n \leqq N$ which interact with their closest neighbours with energy J. An external field H may also act on the spins. The problem is to describe the chain at equilibrium. Mathematically (or physically), the poblem reduces to finding the minimum of the Hamiltonnian (potential)

$$\mathcal{H}(\sigma) = -J\sum_{n=1}^{N-1} \sigma_n \sigma_{n+1} - H\sum_{n=1}^{N} \sigma_n,$$

when $\sigma = (\sigma_1 \sigma_2 \cdots \sigma_N)$ runs through the 2^N positions in the set $\{-1, +1\}^N$.

This is the simplest case and easily solved. Indeed, if $J > 0$ (ferromagnetic case), then

$$\mathcal{H}_{min} = \min \mathcal{H}(\sigma) = -(N-1)J-N|H|$$

is attained if for all n, $\sigma_n = \mathrm{sgn}(H)$. The external field imposes its direction on each one of the sites. This is a reasonable description of a piece of iron in a magnetic field.

If $J < 0$ (paramagnetic case), the situation is more complicated.

If the external field is weak, then

$$\mathcal{H}_{min} = (N-1)J - \frac{1}{2}(1-(-1)^N)|H|.$$

If the external field is strong, then $\mathcal{H}_{min} = (N-1)J - N|H|$.

We propose to discuss the more complex case when the substance is a mixture of ferromagnetic and paramagnetic matter. We are now given a sequence $\varepsilon = (\varepsilon_n) \in \{-1,+1\}^N$, and we wish to study the Ising chain governed by the Hamiltonian:

$$\mathcal{H}_\varepsilon(\sigma) = -J\sum_{n=1}^{N-1} \varepsilon_n \sigma_n \sigma_{n+1} - H\sum_{n=1}^{N} \sigma_n,$$

where we can assume without loss of generality that $J > 0$ (if $J > 0$, change all signs in ε). To find the configuration σ which minimizes $\mathcal{H}_\varepsilon(\sigma)$ is not easy. See Derrida and Gardner [14]. We shall here restrict our study to finding the induced magnetic field at $T=0$.

Solving the model at temperature $T>0$ means to compute the partition function:

$$Z_N(T) = \sum_{\sigma \in \{-1,+1\}^N} \exp(-\beta\,\mathcal{H}_\varepsilon(\sigma)),$$

where $\beta=1/T$. The knowledge of $Z_N(T)$ enables us to calculate all thermodynamic and magnetic properties of the chain. The important quantity is the free energy per spin:

$$\lim_{N\to\infty} \frac{1}{N} \log Z_N(T) = -\frac{1}{T}\psi,$$

and the related functions.

Let Z_n^+ be the partition function where the spin at the nth site is fixed and equal to $\sigma_n = +1$. Similarly, Z_n^- is the partition function where $\sigma_n = -1$. Hence $Z_N(T) = Z_n^+ + Z_n^-$, it can be shown that as N tends to infinity, Z_n^+ and Z_n^- are given by

$$\begin{bmatrix} Z_n^+ \\ Z_n^- \end{bmatrix} = \begin{bmatrix} \prod_{q=0}^{n-1} M_q \end{bmatrix} \begin{bmatrix} Z_0^+ \\ Z_0^- \end{bmatrix},$$

where the matrix M_q is

$$M = \begin{bmatrix} z^{\varepsilon_q + H/J} & z^{-\varepsilon_q + H/J} \\ z^{-\varepsilon_q - H/J} & z^{\varepsilon_q - H/J} \end{bmatrix},$$

with $z = \exp \beta J$.

Let us examine the chain as T approaches to 0 from above. The parameter z tends to infinity so that Z_n^+ and Z_n^- essentially behave as monomials in z:

$$Z_n^+ \text{ is of the order of } z^{a_n},$$

$$Z_n^- \text{ is of the order of } z^{b_n}.$$

The difference:

$$d_n = a_n - b_n = \lim_{T\to 0} \frac{T}{J} \log \frac{Z_n^+}{Z_n^-}$$

is interpreted by physicists as the magnetic field at site n induced by sites $n-1, n-2, \cdots$. The sequence $d = (d_n)$ is seen to be a solution of the recurrence relation (see [2] or [15] for example):

$$d_{n+1} = 2H/J + \varepsilon_n \operatorname{sgn}(d_n) \min\{2, |d_n|\}.$$

The value of d_0 is irrelevant since, for large n, d_n is independent of d_0. The macroscopic description of the chain ignores finite perturbations.

Theorem 4 (Allouche, Mendès France [2],[3]). *If the sequence* $\varepsilon = (\varepsilon_n)$ *is* k-*automatic, then the sequence* $d = (d_n)$ *defined by* (1) *is* k-*automatic.*

The proof we give in our paper [2] is constructive. It shows for example that if (ε_n) is the Thue-Morse sequence on the symbols +,- (see Example 1 Section 2) and if H=J, then the sequence $d = (d_n)$ is generated by a 105 state automaton. There is however a simplification if one considers the shifted sequence (d_{n+2}), $n \geq 0$. It is generated by the following 2-automaton.

A → 0
B → 2
C → 4
D → 2
E → 4

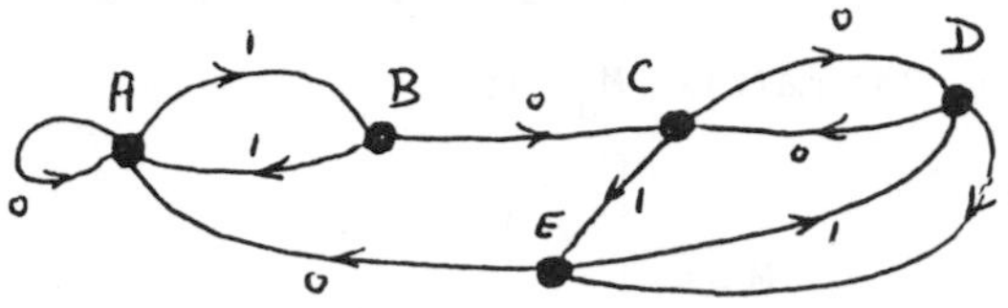

One dimension physics is a very lively domain with some beautiful problems. I have discussed rapidly the Ising chain. Let me end this survey by mentioning a rather frustrating problem. Masses m and M are linked by identical springs along an infinite chain.

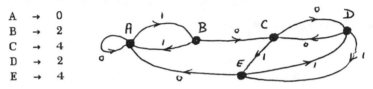

The problem is to describe the vibrations of the system. The problem appears to be very difficult. It is however well understood when m = M or when the masses m and M are either periodically or randomly distributed.

Allouche and Peyriere [4] on the one hand and Axel, Allouche, Kleman, Mendès France and Peyriere [6] on the other hand have obtained some results when the masses m and M are distributed automatically. It would be interesting to have a general theory.

This last Section 6 is incomplete. Many problems remain unsolved even though some are probably not difficult. We hope to fill in some of the gaps in a near future.

REFERENCES

[1] Allouche, J.P.: Automates finis en théorie des nombres, Expositions Mathematicae, 5, 239-266 (1987).

[2] Allouche, J.P. and Mendès France, M.: Quasicrystal Ising chain and automata theory, J. Stat. Phys., 42, 809-821 (1986).

[3] Allouche, J.P. and Mendès France, M.: Finite automata and zero temperature quasicrystal Ising chain, J. de Physique, Colloque C3, Suppl. 7, 47, C3.63-C3.73 (1986).

[4] Allouche, J.P. and Peyriere, J.: Sur une formule de récurrence sur les traces de produits de matrices associés à certaines substitutions, C. R. A. S. Paris, 302, 1135-1136 (1986).

[5] Allouche, J.P., Betrema, J. and Shallit, J.O.: Sur les points fixes de morphismes du monoïde libre, to appear.

[6] Axel, F., Allouche, J.P., Kleman, M., Mendès France, M. and Peyriere, J.: Vibrational modes in a one dimensional quasi-alloy: the Morse case, J. Phys., Colloq. C3, Suppl. 7, 47, C3.181-C3.186 (1986).

[7] Baker, A.: Continued fractions of transcendental numbers, Mathematika, 9, 1-8 (1962).

[8] Baum, L. and Sweet, M.: Continued fractions of algebraic power series in characteristic 2, Ann. of Math., 103, 593-610 (1976).

[9] Baum, L. and Sweet, M.: Badly approximable power series in characteristic 2, Ann. of Math., 105, 573-580 (1977).

[10] Beyer, W.A., Metropolis, N. and Neegard, J.R.: Statistical study of the digits of some square roots of integers various bases, Math. Comp., 24, 455-473 (1970).

[11] Blanchard, A. and Mendès France, M.: Symétrie et transcendance, Bull. Sci. Math., 106, 325-335 (1982).

[12] Christol, G., Kamae, T., Mendès France, M. and Rauzy, G.: Suites algébriques, automates et substitutions, Bull. Soc. Math. France, 108, 401-419 (1980).

[13] Cobham, A.: On the base dependence of sets of numbers recognisable by finite automata, Math. Systems Theory, 3, 186-192 (1969).

[14] Dekking, F.M., Mendès France, M. and Van der Poorten, A.J.: Folds! Mathematical Intelligencer, 4, 130-138; 173-181; 190-195 (1982).

[15] Derrida, B. and Gardner, E.: Metastable states of a spin glass chain at 0 temperature, J. de Physique, 47, 959 (1986).

[16] Ford, J., Chaos, Solving the unsolvable, Predicting the unpredictable! In chaotic dynamics and fractals, edited by M.F. Barnsley, S.G. Demko, Academic Press, London-New York, 1-56, 1986.

[17] Loxton, Van der Poorten, A.J.: Arithmetic properties of the solutions of a class of functional equations, J. Reine. Angew. Math., 330, 159-172 (1982).

[18] Mahler, K.: Ein Annalogen zu einem Schneiderschen Satz, Proc. Akad. Wetensch. Amsterdam, 39, 633-640, 729-737 (1936).

[19] Mendès France, M.: The Ising chain with nonconstant external field, J. Stat. Phys., 45, 89-97 (1986).

[20] Mills, W.H. and Robbins, D.P.: Continued fractions for certain algebraic power series, J. Number Theory, 23, 388-404.

[21] Shallit, J.O.: Simple continued fractions for some irrational numbers, J. Number Theory, 2, 209-217 (1979).

Proc. Prospects
of Math. Sci.
World Sci. Pub.
141-156, 1988

AN EXPLICIT FORMULA FOR THE AVERAGE OF SOME q-ADDITIVE FUNCTIONS

LEO MURATA

Meiji Gakuin University
Department of Mathematics
Shiroganedai, Minato-ku
Tokyo 108
Japan

JEAN-LOUP MAUCLAIRE

U.E.R. de Math. et Inf.
Université Paris VII
2, Place Jussieu
75251 Paris
France

1. INTRODUCTION AND RESULTS

Let q be an arbitrary fixed natural number $\geqq 2$, then a natural number n can be written in a unique way as

$$n = \sum_{k=0}^{\infty} a_k(n)\, q^k \ , \ 0 \leqq a_k(n) \leqq q-1 \ ,$$

which is called the q-adic expansion of n.

An arithmetical function g(n) is called q-additive, if g(n) satisfies the relations:

$$g(0) = 0, \quad g(n) = \sum_{k=0}^{\infty} g(a_k(n)q^k) \ ,$$

whenever n has the above q-adic expansion (cf. Gelfond [1]). When we give the values of the function g on the set $\{rq^k : 1 \leqq r \leqq q-1, k \in \mathbb{N} \cup \{0\}\}$, then the q-additive function g(n) is determined uniquely by the above relations, and vice versa. The most famous and typical example of a q-additive function is the function "sum of digits", $S_q(n)$, which is defined by $S_q(n) = \sum_{k=0}^{\infty} a_k(n)$. We have known many results concerning an asymptotic average of $S_q(n)$ (cf. Delange [2], Introduction), and in 1975, Delange proved a definitive theorem in the above cited article:

Theorem 1. (Delange, [2]) *We have, for any* $N \in \mathbb{N}$, *that*

$$\frac{1}{N}\sum_{n=0}^{N-1} S_q(n) = \frac{q-1}{2\log q}\log N + F(\frac{\log N}{\log q}) , \tag{1}$$

where the function $F(x)$ *is a periodic function with period* 1, *defined by either of the following two ways:*

(I) $$F(x) = \frac{q-1}{2}(1+[x]-x)$$

$$+ q^{1+[x]-x}\sum_{r=0}^{\infty} q^{-r}\int_0^{q^r(q^{-1-[x]+x})}([qt] - q[t] - \frac{q-1}{2})\, dt ,$$

where $[x]$ *denotes the largest integer not exceeding* x.

(II) $$F(x) = \sum_{k\in\mathbb{Z}} C_k \exp(2\pi ikx),$$

that is the Fourier expansion of $F(x)$, *where*

$$C_0 = \frac{q-1}{2\log q}(\log(2\pi) - 1) - \frac{q+1}{4} ,$$

$$C_k = i\,\frac{q-1}{2\pi k}(1 + \frac{2\pi ik}{\log q})^{-1}\zeta(\frac{2\pi ik}{\log q}) , \quad k \neq 0 ,$$

and $\zeta(s)$ *denotes the Riemann zeta function.*

In this article, we will develop our proof to obtain the following rather general theorem, which contains the formulas (1) and (II). In order to give our main result, we have to put some assumptions and to introduce some notations: Let $g(n)$ be a q-additive function.

Assumption (H) $\left\{\begin{array}{l}\text{Suppose that each of the series} \\ \sum_{k=0}^{\infty} g(rq^k)x^{-(k+1)} = f_r(x),\ 1 \leqq r \leqq q-1, \\ \text{converges for any } |x| \text{ sufficiently large enough,} \\ \text{and represents a rational function.}\end{array}\right.$

We consider, for $1 \leqq r \leqq q-1$, the finite sets

$$\Pi_r = \{\ P\ ;\ P \text{ is a pole of } f_r(x)\ \} ,$$

and for every $P \in \Pi_r$, we define the numbers $d_{r,p}$, $\rho(P)$, $C_{r,P,\ell}$, $D_{r,\rho}(i)$ as follows:

$d_{r,P}$ = the order of the pole of $f_r(x)$ at $x = P$,

$\rho = \rho(P)$ is the complex number which satisfies

$$q^{\rho(P)} = P \quad \text{and} \quad 0 \leqq \mathrm{Im}(\rho(P)) < \frac{2\pi}{\log q} ,$$

and, of course, this $\rho(P)$ can always be determined uniquely, and

$$f_r(q^s) = \sum_{P\in\Pi_r} \sum_{\ell=1}^{d_{r,P}} C_{r,P,\ell} (q^s - q^{\rho(P)})^{-\ell}, \quad C_{r,P,\ell} \in \mathbf{C} \tag{2}$$

$$= \sum_{P\in\Pi_r} \sum_{\ell=1}^{d_{r,P}} C_{r,P,\ell} \sum_{i=1}^{\infty} D_{r,\rho}(i) (s - \rho(P))^{i-\ell}, \quad D_{r,\rho}(i) \in \mathbf{C} .$$

In addition we make use of

$$\zeta(s,a) = \sum_{n=0}^{\infty} (n + a)^{-s} , \quad 0 < a \leqq 1, \text{ the Hurwitz zeta function,}$$

$$\Phi_r(s) = \zeta(s,\frac{r}{q}) - \zeta(s,\frac{r+1}{q}) , \quad \Phi_r^{(h)}(s) = \frac{d^h}{(ds)^h} \Phi_r(s), \quad h\in\mathbb{N} .$$

Our main theorem is:

<u>Theorem 2</u>. *If every* $P\in \bigcup_{r=1}^{q-1} \Pi_r$ *satisfies the condition* $|P| > q^{-\frac{1}{2}}$, *then we have the following explicit formulas for any* $n\in\mathbb{N}$, *under the assumption* (H) *and with the above notations:*

$$\frac{1}{N} \sum_{n=0}^{N-1} g(n) = \sum_{r=1}^{q-1} \sum_{\ell=1}^{d_{r,1}} \frac{1}{q(\ell!)} C_{r,1,\ell} D_{r,0}(0) (\log N)^{\ell} \delta_{P,1}$$

$$+ \sum_{r=1}^{q-1} \sum_{P\in\Pi_r} \sum_{\ell=1}^{d_{r,P}} \left\{ \sum_{t=0}^{\ell-1} C_{r,P,\ell} N^{\rho(P)} (\log N)^t F_{r,P,\ell,t}(\frac{\log N}{\log q}) \right.$$

$$\left. + \frac{1}{q} C_{r,P,\ell} (\delta_{P,1} - 1)^{2\ell} (1 - P)^{-\ell} \right\} , \tag{3}$$

where $\delta_{P,1}$ *denotes Kronecker's* δ*-function. Furthermore, every function* $F_{r,P,\ell,t}(x)$ *is a periodic function with period* 1 *and its Fourier coefficients are given explicitly as follows:*

$$F_{r,P,\ell,t}(x) = \sum_{k\in\mathbf{Z}} A_{r,P,\ell,t}(k)\exp(2\pi ikx) ,$$

with

$$A_{r,P,\ell,t}(k) = \begin{cases} \frac{1}{t!} \sum_{\substack{i+j+h=\ell-t \\ i,j,h\geq 0}} (-1)^j D_{r,0}(i) \frac{1}{h!} \Phi_r^{(h)}(0) , \\ \qquad\qquad\qquad \text{if } P = 1 \text{ and } k = 0 , \\ \frac{1}{t!} \sum_{\substack{i+j+h=\ell-t-1 \\ i,j,h\geq 0}} (-1)^j D_{r,\rho}(i) \frac{1}{h!} \Phi_r^{(h)}(\rho_k) \\ \times \left\{(\rho_k)^{-j-1} - (\rho_k+1)^{-j-1}\right\} , \quad \text{if } P \neq 1 \text{ or } k \neq 0 , \end{cases}$$

where $\rho_k = \rho(P) + \frac{2\pi i}{\log q} k$.

Here we remark the link between our result and above cited Delange's result. Since the function "sum of digits", $S_q(n)$, is defined by $S_q(rq^k) = r$ for any $k\in\mathbb{N}$, we know that

$$f_r(x) = r(x - 1)^{-1} ; \quad \Pi_r = \{1\} ; \quad d_{r,1} = 1 \quad \text{for any } r.$$

Then,

$$C_{r,1,1} = r, \ \rho(P) = 0, \ D_{r,0}(0) = \frac{1}{\log q} , \ D_{r,0}(1) = -\frac{1}{2} ,$$

and

$$A_{r,1,1,0}(k) = \begin{cases} \frac{1}{\log q} \Phi_r(\frac{2\pi ik}{\log q}) \left((\frac{2\pi ik}{\log q})^{-1} - (\frac{2\pi ik}{\log q} + 1)^{-1}\right) , \\ \qquad\qquad\qquad \text{if } k \neq 0 , \\ -\frac{1}{2} \Phi_r(0) + (-1) \frac{1}{\log q} \Phi_r(0) + \frac{1}{\log q} \Phi_r^{(1)}(0) , \\ \qquad\qquad\qquad \text{if } k = 0 . \end{cases}$$

By making use of the relations

$$\sum_{r=1}^{q-1} r \Phi_r(s) = (q^s - q) \zeta(s) ,$$

$$\sum_{r=1}^{q-1} r \Phi_r^{(1)}(s) = q^s(\log q) \zeta(s) + (q^s - q) \zeta^{(1)}(s) ,$$

$$\zeta^{(1)}(0) = -\frac{\log 2\pi}{2} ,$$

our theorem gives the explicit formula (1) and (II).

The authors already announced this result in [3] without proof. We are very grateful to Professor H. Delange for his kind comments to the original manuscript.

2. PRELIMINARY LEMMAS

We shall prove in this Section a functional equation which is closely related to a q-additive function, and prepare some Lemmas.

Lemma 1. *Let* $g(n)$ *be a* q-*additive function, and we suppose that for* $r = 1, 2, \cdots, q-1$, *the series* $\sum_{k=0}^{\infty} g(rq^k)\, x^{-(k+1)}$ *converges absolutely for any* $|x| > Z_r$, *where* Z_r *is some positive constant and the series represents an analytic function* $f_r(x)$ *in this domain. Then we have the functional equation*

$$\sum_{n=1}^{\infty} (g(n) - g(n-1))\, n^{-s} = \sum_{r=1}^{q-1} \Phi_r(s)\, f_r(q^s)\,, \tag{4}$$

for any s *satisfying* $|q^s| > Z_0 = \max_{1 \leq r \leq q-1} \{Z_r\}$.

Proof. (The following elegant proof is due to Prof. Delange.)

Any natural number n can be written by $n = rq^u + mq^{u+1}$ in the unique way, where $1 \leq r \leq q-1$ and $u, m \in \mathbb{N} \cup \{0\}$. Then

$$g(n) - g(n-1) = g(rq^u) - g((r-1)q^u) - \sum_{h \leqq u-1} g((q-1)q^h)$$

and

$$\sum_{n=1}^{\infty} \frac{g(n)-g(n-1)}{n^s} = \sum_{r=1}^{q-1} \sum_{u=0}^{\infty} \sum_{m=0}^{\infty} \frac{g(rq^u)-g((r-1)q^u) - \sum_{h \leqq u-1} g((q-1)q^h)}{(rq^u+mq^{u+1})^s}$$

$$= \sum_{r=1}^{q-1} \sum_{u=0}^{\infty} \left\{ g(rq^u)-g((r-1)q^u) - \sum_{h \leqq u-1} g((q-1)q^h) \right\} q^{-(u+1)s}\, \zeta(s, \tfrac{r}{q})$$

$$= \sum_{r=1}^{q-1} \zeta(s, \tfrac{r}{q}) \left\{ f_r(q^s) - f_{r-1}(q^s) - \sum_{u=0}^{\infty} g((q-1)q^u) \sum_{t \geqq u+2} q^{-ts} \right\}$$

$$= \sum_{r=1}^{q-1} \zeta(s, \tfrac{r}{q}) \left\{ f_r(q^s) - f_{r-1}(q^s) - \frac{q^{-s}}{1-q^{-s}}\, f_{q-1}(q^s) \right\}.$$

Since $f_0(q^s) = 0$ and $\sum_{r=1}^{q-1} \zeta(s,\frac{r}{q}) = (q^s - 1)\,\zeta(s)$, we obtain our functional equation.

< Q.E.D. >

We quote here the Perron formula without proof.

Lemma 2. *Let* $\{\mu_n\}_{n=1}^{\infty}$ *be a sequence of positive numbers satisfying* $\mu_{n+1} \geqq \mu_n$ *for all* n, *and* $\lim_{n\to\infty} \mu_n = \infty$. *We assume that the Dirichlet series* $\sum_{n=1}^{\infty} a_n(\mu_n)^{-s}$ *converges absolutely for* $\mathrm{Re}(s) = C > 0$, *and denote the sum by* $f(s)$. *Then we have, for any* $x > 0$,

$$\int_0^x A(t)\,dt = \frac{1}{2\pi i}\int_{C-i\infty}^{C+i\infty} f(s)\,\frac{x^{s+1}}{s(s+1)}\,ds\,, \tag{5}$$

where $A(t) = \sum_{n\in\{n;\mu_n\leqq t\}} a_n$.

Now we put $a_n = g(n) - g(n-1)$ and $\mu_n = n$, then the above two Lemmas give the following:

Lemma 3. *Under the hypothesis of* Lemma 1, *we have*

$$\frac{1}{N}\sum_{n=1}^{N-1} g(n) = \sum_{r=1}^{q-1} \frac{1}{2\pi i}\int_{C-i\infty}^{C+i\infty} \Phi_r(s)\,\frac{f_r(q^s)}{s(s+1)}\,N^s\,ds\,, \tag{6}$$

for any positive integer N, *where* C *is a real number satisfying* $q^C > Z_0$.

Proof. We put $x = N$ in (5); then, Lemma 1 gives

$$\frac{1}{2\pi i}\int_{C-i\infty}^{C+i\infty} \sum_{r=1}^{q-1} \Phi_r(s)\,\frac{f_r(q^s)}{s(s+1)}\,N^{s+1}\,ds$$

$$= \int_0^N \Big(\sum_{n\leqq t} (g(n) - g(n-1))\Big)\,dt = \sum_{n=0}^{N-1} g(n)\,.$$

Multiplying by N^{-1} the both sides, we get (6).

< Q.E.D. >

This shows that the problem reduces to the calculation of a formula of the type

$$\frac{1}{2\pi i}\int_{C-i\infty}^{C+i\infty}\Phi_r(s)\,\frac{f_r(q^s)}{s(s+1)}\,N^s ds\ ,\quad N\in\mathbb{N}\ ,$$

and in order to compute this integral, we have to prepare some Lemmas concerning the Hurwitz zeta function.

It is well-known (cf. Titchmarsh [4], Chap. 2) that $\zeta(s,a)$ is meromorphic in the whole s-plane, has a simple pole at $s = 1$,

$$\operatorname*{Res}_{s=1}\zeta(s,a) = 1\ , \tag{7}$$

and

$$\zeta(0,a) = \frac{1}{2} - a\ . \tag{8}$$

Furthermore, we have:

Lemma 4. *Let* β *be some positive constant, then for any* $\sigma > -\beta$ *, we have uniformly*

$$\Phi_r^{(h)}(\sigma+it) = O(|t|^{\frac{1}{2}+\beta+\varepsilon})\ ,$$

for any small $\varepsilon > 0$, *where the constant implied by O-symbol depends only on* β, h *and* ε .

Proof. We notice first that, once we proved

$$\Phi_r(\sigma+it) = O(|t|^{\frac{1}{2}+\beta}),\quad \text{for any}\ \ \sigma > -\beta\ , \tag{9}$$

then we can easily obtain our Lemma. In fact, we have

$$\max_{|s-(\sigma+it)|\leq\varepsilon}|\Phi_r(s)| = O((|t|+\varepsilon)^{\frac{1}{2}+\beta+\varepsilon}) = O(|t|^{\frac{1}{2}+\beta+\varepsilon})$$

and, making use of this estimate on the circle $|s-(\sigma+it)| = \varepsilon$, Cauchy's estimates for derivatives of analytic function give our conclusion.

Now we suppose $\alpha > 0$. If $\sigma \geq 1+\alpha$, then (9) is true. So it remains to prove (9) only for a $\sigma+it$ satisfying $-\beta \leq \sigma \leq 1+\alpha$ and $|t| \geq 1$. We can prove this from Phragmén-Lindelöf theorem. It is known that $\zeta(s,a)$ is a function of finite order and of course we have

$$\Phi_r(1+\alpha+it) = O(|t|^{\varepsilon}) , \tag{10}$$

where ε is an arbitrary positive number. For any $\sigma < 0$, Hurwitz formula shows, for $s = \sigma+it$, that

$$\zeta(s,a) = \frac{2\Gamma(1-s)}{(2\pi)^{1-s}} \left\{ \sin(\frac{\pi s}{2}) \sum_{n=1}^{\infty} \frac{\cos(2n\pi a)}{n^{1-s}} + \cos(\frac{\pi s}{2}) \sum_{n=1}^{\infty} \frac{\sin(2n\pi a)}{n^{1-s}} \right\}$$

(cf. [4], formula (2.17.3)), and from here we can easily obtain an estimate:

$$|\zeta(s,a)| \leq 2^{\sigma+1} \pi^{\sigma} \exp(\frac{\pi}{2}|t|) \zeta(1-\sigma) |\Gamma(1-s)| .$$

Since $|\Gamma(1-\sigma-it)| \leq C_1 |t|^{1-\sigma-\frac{1}{2}} \exp(-\frac{\pi}{2}|t|)$, we get

$$\Phi_r(-\beta+it) = O(|t|^{\frac{1}{2}+\beta}) . \tag{11}$$

From (10) and (11), Phragmén-Lindelöf theorem proves our estimate (9).

< Q.E.D. >

Lemma 5. *For any positive* γ *satisfying* $-\frac{1}{2} < -\gamma < 0$, *and for any* a *satisfying* $0 < a \leq 1$, *we have*

$$\frac{1}{2\pi i} \int_{-\gamma-i\infty}^{-\gamma+i\infty} \frac{\zeta(s,a)}{s(s+1)} x^s \, ds$$

$$= a + \frac{1}{2x} \{([x-a]+2)([x-a]-1) - (x+2)(x-1)\}$$

$$+ \frac{[x-a]+1}{x} \{(x-a) - [x-a]\} .$$

Especially, this is equal to zero, when x *is a positive integer.*

Proof. We define $\lambda(t) = [t-a] + 1$ and, for $n \in \mathbb{N}$, put $a_n = \lambda(n) - \lambda(n-1)$, $\mu_n = n + a$. Then $a_n = 1$ for any $n \geq 0$ and it follows from our Lemma 2 that

$$\Lambda(x) = \frac{1}{x} \int_0^x \lambda(t) \, dt = \frac{1}{2\pi i} \int_{C-i\infty}^{C+i\infty} \zeta(s,a) \frac{x^s}{s(s+1)} \, ds , \tag{12}$$

where C is a constant > 1. Now, from (7) and (8), we know that the integrand, $\zeta(s,a)x^s s^{-1}(s+1)^{-1}$, has three singularities:

simple pole at $s = 1$ with a residue $\frac{1}{2}x$,

simple pole at $s = 0$ with a residue $\frac{1}{2}-a$,

simple pole at $s = -1$.

We want to shift here the contour of integral in (12) to the line $\mathrm{Re}(s) = -\gamma$. For this purpose, it is sufficient to prove that

$$\lim_{N\to\infty} \left| \int_{C+iN}^{-\gamma+iN} \zeta(s,a) \frac{x^s}{s(s+1)} \, ds \right| = 0 ,$$

and this can be proved easily from the estimate

$$\zeta(s,a) = O(|t|^{\frac{1}{2}+\gamma}) ,$$

which is uniform in s whenever $-\gamma \leqq \mathrm{Re}(s) \leqq 0$; this estimate is obtained in Lemma 4. Then Cauchy's integral theorem gives

$$\frac{1}{2\pi i} \int_{-\gamma-i\infty}^{-\gamma+i\infty} \zeta(s,a) \frac{x^s}{s(s+1)} \, ds = \Lambda(x) + (-\frac{1}{2}x) + (-\frac{1}{2} + a) .$$

On the other hand,

$$\begin{aligned} \Lambda(x) &= \frac{1}{x} \int_0^x \lambda(t) \, dt \\ &= \frac{1}{x} \{1 + \sum_{i=2}^{[x-a]} i + (x - [x-a] - a)([x-a] + 1)\} \\ &= \frac{1}{x} + \frac{1}{2x}([x-a] + 2)([x-a] - 1) + \frac{[x-a]+1}{x}((x-a) - [x-a]) . \end{aligned}$$

This proves the formula in our Lemma 5.

Now we take x as a positive integer m , then $[x-a] = m - 1$ and it is easy to verify our assertion.

< Q.E.D. >

3. PROOF OF THEOREM 2

In this Section, we assume that the q-additive function $g(n)$ satisfies the assumption (H) introduced in Section 1, i.e.

$$\sum_{k=0}^{\infty} g(rq^k)\, x^{-(k+1)} = f_r(x) = \frac{G_r(x)}{H_r(x)} \; ; \; G_r(x),\, H_r(x) \in \mathbb{C}[x] \text{ *)}$$

for $1 \leqq r \leqq q-1$. We use the notations Π_r, $d_{r,P}$ and so on defined in Section 1 also. We decomposed every $f_r(x)$ into

$$f_r(x) = \sum_{P\in\Pi_r} \sum_{\ell=1}^{d_{r,P}} C_{r,P,\ell}(x - P)^{-\ell}, \quad C_{r,P,\ell} \in \mathbb{C} \,, \tag{13}$$

(cf. (2)). Furthermore, if $|x|>|P|$, every function $C_{r,P,\ell}(x - P)^{-\ell}$ can be expanded into

$$C_{r,P,\ell}(x - P)^{-\ell} = C_{r,P,\ell} \sum_{k=0}^{\infty} \binom{k+\ell-1}{k} P^k\, x^{-(\ell+k)} \,.$$

Consequently, combining (13) with Lemma 3, we have

$$\frac{1}{N} \sum_{n=1}^{N-1} g(n)$$

$$= \sum_{r=1}^{q-1} \sum_{P\in\Pi_r} \sum_{\ell=1}^{d_{r,P}} C_{r,P,\ell}\, \frac{1}{2\pi i} \int_{C-i\infty}^{C+i\infty} \Phi_r(s)\, \frac{N^s}{s(s+1)}\, (q^s - P)^{-\ell}\, ds \,. \tag{14}$$

From now on, we assume that $|P| > q^{-\frac{1}{2}}$. Our purpose is to calculate the definite integral

$$\frac{1}{2\pi i} \int_{C-i\infty}^{C+i\infty} \Phi_r(s)\, \frac{N^s}{s(s+1)}\, (q^s - P)^{-\ell}\, ds \,. \tag{15}$$

In this Section we abbreviate $\rho = \rho(P)$, that is

$$P = q^{\rho} \,, \quad 0 \leqq \mathrm{Im}(\rho) < \frac{2\pi}{\log q} \,, \; -\frac{1}{2} < \mathrm{Re}(\rho) \,.$$

If $P = q^{\rho}$ is a pole of the function $f(x)$, then every $\rho + \frac{2\pi i}{\log q} k$, $k \in \mathbf{Z}$ is also a pole of the function $f(q^s)$. So we put

$$\rho_k = \rho + \frac{2\pi i}{\log q} k \,, \quad k \in \mathbf{Z} \,,$$

---

*) It is clear that $\deg H_r(x) \geqq \deg G_r(x) + 1$.

and

$$R_{r,\rho}(\rho_k) = \operatorname*{Res}_{s=\rho_k} \Phi_r(s) \frac{N^s}{s(s+1)} (q^s - q^\rho)^{-\ell} .$$

i) In the case of $\rho_k \neq 0$, i.e. either "$\rho \neq 0$" or "$\rho = 0$ and $k \neq 0$", we can expand the integrand in (15) at $s = \rho_k$ as follows:

$$(q^s - q^\rho)^{-\ell} = \sum_{i=0}^{\infty} D_{r,\rho}(i)\ (s - \rho_k)^{i-\ell} ,$$

$$\frac{1}{s(s+1)} = \sum_{j=0}^{\infty} \left\{(-1)^j \left((\rho_k)^{-j-1} - (\rho_k+1)^{-j-1}\right)\right\} (s - \rho_k)^j ,$$

$$\Phi_r(s) = \sum_{h=0}^{\infty} \left(\frac{1}{h!} \Phi_r^{(h)}(\rho_k)\right) (s - \rho_k)^h ,$$

$$N^s = \sum_{t=0}^{\infty} \left\{\frac{1}{t!} N^\rho (\log N)^t \exp(2\pi ik \frac{\log N}{\log q})\right\} (s - \rho_k)^t .$$

Therefore,

$$R_{r,\rho}(\rho_k) = \sum_{t=0}^{\ell-1} N^\rho (\log N)^t A_{r,P,\ell,t}(k) \exp(2\pi ik \frac{\log N}{\log q}) ,$$

with

$$A_{r,P,\ell,t}(k) = \frac{1}{t!} \sum_{\substack{i+j+h=\ell-t-1 \\ i,j,h \geq 0}} (-1)^j D_{r,\rho}(i) \frac{1}{h!} \Phi_r^{(h)}(\rho_k) \times \left((\rho_k)^{-j-1} - (\rho_k+1)^{-j-1}\right) .$$

ii) In the case of $\rho_k = 0$, i.e. $\rho = 0$ and $k = 0$, the Laurent expansion of $s^{-1}(s+1)^{-1}$ changes into

$$\frac{1}{s(s+1)} = \sum_{j=0}^{\infty} (-1)^j s^{j-1} ,$$

and correspondingly

$$R_{r,\rho}(0) = \frac{1}{q} \frac{1}{\ell!} D_{r,0}(0) (\log N)^\ell + \sum_{t=0}^{\ell-1} (\log N)^t A_{r,1,\ell,t}(0) ,$$

with

$$A_{r,1,\ell,t}(0) = \frac{1}{t!} \sum_{\substack{i+j+h=\ell-t \\ i,j,h\geqq 0}} (-1)^j\, D_{r,\rho}(i)\, \frac{1}{h!}\, \Phi_r^{(h)}(0) .$$

iii) In the case of $\rho \neq 0$, the point $s = 0$ is a simple pole of the integrand, and we have

$$R_{r,\rho}(0) = \frac{1}{q}\,(1 - q^{\rho})^{-\ell} .$$

Now we take a positive constant β satisfying $-\frac{1}{2} \geqq -\beta < \mathrm{Re}(\rho)$, and consider, for any u in $\mathbb{N}$, the contour Γ_u defined by the following figure Fig. 1:

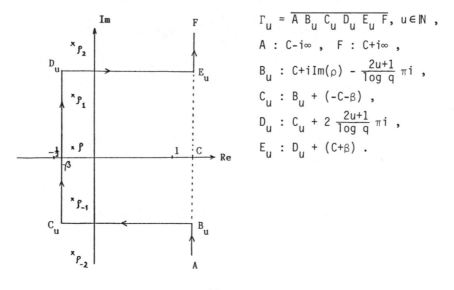

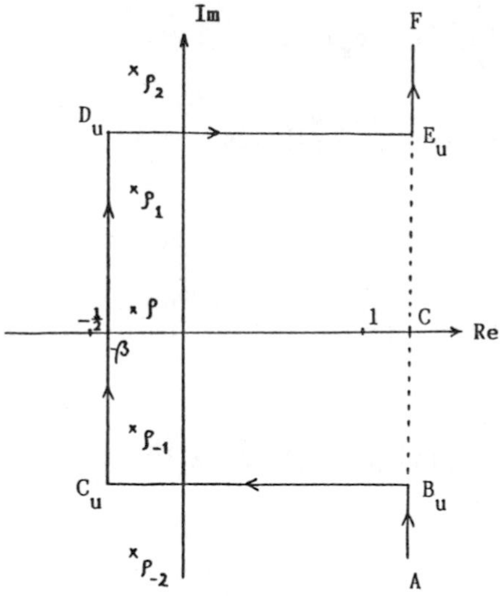

$\Gamma_u = \overline{A\ B_u\ C_u\ D_u\ E_u\ F}$, $u \in \mathbb{N}$,

$A : C - i\infty$, $F : C + i\infty$,

$B_u : C + i\,\mathrm{Im}(\rho) - \dfrac{2u+1}{\log q}\pi i$,

$C_u : B_u + (-C-\beta)$,

$D_u : C_u + 2\,\dfrac{2u+1}{\log q}\pi i$,

$E_u : D_u + (C+\beta)$.

Fig. 1

Now, we shift the contour of the integral in (15) to the contour Γ_u , $u = 1, 2, \cdots$, one by one, and consider this integral when u tends to infinity. Concerning the residues of the integrand, Lemma 4 gives the estimate:

$$\begin{aligned}\sum_{k\in\mathbb{Z}} A_{r,P,\ell,t}(k) &= O\left(\sum_{k\in\mathbb{Z}} \left(|(\rho_k)^{-1} - (\rho_k+1)^{-1}| \times |\Phi_r^{(h)}(\rho_k)|\right)\right) \\ &= O\left(\int_1^{\infty} k^{-2/3+\mathrm{Min}(0,-\mathrm{Re}(\rho))+\varepsilon}\, dk\right) , \qquad (16)\end{aligned}$$

for any $\varepsilon > 0$, and this integral converges. In addition, Lemma 4 gives also

$$\int_{\overline{B_u C_u}} |\Phi_r(s) \frac{N^s}{s(s+1)} (q^s - q^\rho)^{-\ell}| \, ds = O(u^{\frac{1}{2}+\beta-2}) < \infty .$$

This shows that

$$(15) = \sum_{k\in\mathbf{Z}} R_{r,\rho}(\rho_k) + \frac{1}{2\pi i}\int_{-\beta-i\infty}^{-\beta+i\infty} \Phi_r(s) \frac{N^s}{s(s+1)} (q^s - q^\rho)^{-\ell} \, ds$$
$$+ (\delta_{P,1} - 1)^{2\ell} R_{r,\rho}(0) . \qquad (17)$$

The first term in this right hand side converges absolutely, and

$$\sum_{k\in\mathbf{Z}} R_{r,\rho}(\rho_k) = \delta_{P,1} \frac{D_{r,0}(0)}{q(\ell!)} (\log N)^\ell$$
$$+ \sum_{t=0}^{\ell-1} N^\rho (\log N)^t F_{r,P,\ell,t}(\frac{\log N}{\log q}) ,$$

with

$$F_{r,P,\ell,t}(x) = \sum_{k\in\mathbf{Z}} A_{r,P,\ell,t}(k) \exp(2\pi ikx) .$$

Finally, we calculate the second term in the right hand side of (17). On the line $\mathrm{Re}(s) = -\beta$, it is easy to see that we have the expansion

$$(q^s - q^\rho)^{-\ell} = \sum_{n=0}^{\infty} e_r(n) q^{ns} , \quad e_r(n) \in \mathbf{C} ,$$

and Lebesgue bounded convergence theorem shows that the second term of (17) is equal to

$$\sum_{n=0}^{\infty} e_r(n) \frac{1}{2\pi i} \int_{-\beta-i\infty}^{-\beta+i\infty} \Phi_r(s) \frac{(Nq^n)^s}{s(s+1)} \, ds .$$

Since $Nq^s \in \mathbb{N}$, Lemma 5 says that every term in this formula is equal to zero, and consequently, the second integral term of (17) vanishes. So, taking account of (14), we can obtain our Theorem 2.

< Q.E.D. >

4. APPLICATIONS

In this Section, we present two applications of Theorem 2 to give the explicit average of q-additive functions.

Example 1. Let q be an arbitrary fixed natural number ≥ 2, and $g(n)$ be a q-additive function defined in the following way:

$$g(0) = 0, \quad \text{and} \quad g(rq^k) = C_r(D_r)^k \quad \text{for any} \quad k \in \mathbb{N},$$

where every C_r, D_r $(1 \leq r \leq q-1)$ is a constant satisfying $D_r \neq 1$ and $|D_r| > q^{-\frac{1}{2}}$ for any r. Then Theorem 2 gives that

$$\frac{1}{N} \sum_{n=0}^{N-1} g(n) = \frac{1}{q} \sum_{r=1}^{q-1} \frac{C_r}{1-D_r} + \sum_{r=1}^{q-1} \frac{C_r}{D_r(\log q)} N^{d_r} \times F_r\left(\frac{\log N}{\log q}\right),$$

where d_r is defined by the relations: $|q^{d_r}| = |D_r|$ and $0 \leq \mathrm{Im}(d_r) < \frac{2\pi}{\log q}$, and that every $F_r(x)$ is a periodic function with period 1, whose Fourier expansion is given as follows:

$$F_r(x) = \sum_{k\, Z} C_r(k) \exp(2\pi i k x),$$

$$C_r(k) = \Phi_r\left(d_r + \frac{2\pi i k}{\log q}\right) \left(d_r + \frac{2\pi i k}{\log q}\right)^{-1} \left(d_r + 1 + \frac{2\pi i k}{\log q}\right)^{-1}.$$

Example 2. Let $g(n)$ be a 2-additive function defined by $g(2^k) = k$ for any $k \in \mathbb{N}$. Then $f_1(x) = (x-1)^{-2}$, and $\Phi_1(s) = \zeta(s,\frac{1}{2}) - \zeta(s) = (2^s - 2)\,\zeta(s)$. Then Theorem 2 gives the formula:

$$\frac{1}{N} \sum_{n=0}^{N-1} g(n) = \frac{1}{4}(\log 2)^{-2}(\log N)^2 + (\log N)\, F_1\left(\frac{\log N}{\log 2}\right) + F_2\left(\frac{\log N}{\log 2}\right),$$

where $F_1(x)$ and $F_2(x)$ are periodic functions with period 1, and whose Fourier coefficients are given as follows:

$$F_1(x) = \sum_{k\in \mathbf{Z}} C_k \exp(2\pi i k x)\ ; \quad F_2(x) = \sum_{k\in \mathbf{Z}} D_k \exp(2\pi i k x),$$

$$C_0 = \frac{1}{2}(\log 2)^{-2}(\log(2\pi) - 1) - (\log 2)^{-1},$$

$$C_k = (-2\pi i k(\log 2))^{-1}\, \zeta\left(\frac{2\pi i k}{\log 2}\right) \times \left(1 + \frac{2\pi i k}{\log 2}\right)^{-1}, \quad k \neq 0,$$

$$D_0 = \frac{11}{24} - (\log 2)^{-2} \frac{1}{2} \zeta''(0) - (\log 2)^{-1}(\log(2\pi) - 1)(1 + \frac{1}{2\log 2}) ,$$

$$D_k = (2\pi ik(\log 2))^{-1} (1 + \frac{2\pi ik}{\log 2})^{-1}$$
$$\times \left\{ \left(2 \log 2 + \frac{\log 2}{2\pi ik} + (1 + \frac{2\pi ik}{\log 2})^{-1}\right) \zeta(\frac{2\pi ik}{\log 2}) - \zeta'(\frac{2\pi ik}{\log 2})\right\},$$
$$k \neq 0 .$$

Remark 1. If the function $f(x) = \sum_{k=0}^{\infty} a_k x^{-(k+1)}$ satisfies the following three conditions (A), (B) and (C), then it can be proved that $f(x)$ is a rational function tending to 0 when $|x|$ tends to infinity, i.e.

$$f(x) = \frac{G(x)}{H(x)} ; \quad G(x), \ H(x) \in \mathbf{C}[x] \ \text{ and } \ \deg G(x) < \deg H(x).$$

(A) $f(x)$ is meromorphic in the whole x-plane except $x = 0$,

(B) there exist two real numbers η , ξ such that all poles of $f(x)$ lie in the ring $\eta \leqq |x| \leqq \xi$,

(C) there exists a real number τ such that $f(x) = \sum_{n=0}^{\infty} e_n x^n$ in the domain $|x| < \tau$.

And as we have seen in Section 3, our method needs these three conditions for every $f_r(x)$.

Remark 2. Theorem 2 treats only such a case that "$|P| > q^{-\frac{1}{2}}$ for any $P \in \bigcup_{r=1}^{q-1} \Pi_r$" . We can prove

Lemma 6. *Let* γ *be an arbitrary positive number, and* a *be any real number with* $0 < a \leqq 1$. *Then, for any natural number* n, *we have*

$$\frac{1}{2\pi i} \int_{-\gamma-i\infty}^{-\gamma+i\infty} \Phi_r(s) \frac{n^s}{s(s+1)} \ ds = 0, \quad \text{(cf. Lemma 5).}$$

But we cannot conclude from this Lemma that

$$\frac{1}{2\pi i} \int_{-\gamma-i\infty}^{-\gamma+i\infty} \Phi_r(s) \frac{N^s}{s(s+1)} (q^s - q^\rho)^{-\ell} \ ds = 0 ,$$

because it remains a possibility that the series $\sum_{k\in\mathbf{Z}} R_{r,\rho}(\rho_k)$ does not converge. And it is still open to obtain the explicit formula for q-additive function $g(n)$ without the condition "$|P| > q^{-\frac{1}{2}}$" .

REFERENCES

[1] Gelfond, A.O.: Sur les nombres qui ont des propriétés additives et multiplicatives données, Acta Arith., 13, 259-265 (1967/8).

[2] Delange, H.: Sur la fonction sommatoire de la fonction "somme des chiffres", L'Enseignement Math., 21, 31-47 (1975).

[3] Mauclaire, J-L. and Murata, L.: On q-additive functions, II, Proc. Japan Acad., 59, Ser. A, 441-444 (1983).

[4] Titchmarsh, E.C., The Theory of the Riemann Zeta Function, Oxford Univ. Press, Oxford, 1951.

Proc. Prospects
of Math. Sci.
World Sci. Pub.
157-171, 1988

ASYMPTOTIC DISTRIBUTION AND INDEPENDENCE OF SEQUENCES OF g-ADIC INTEGERS

KENJI NAGASAKA
University of the Air
2-11, Wakaba, Chiba-shi
260, Chiba
Japan

JAU-SHYONG SHIUE
Department of Mathematical Sciences
University of Nevada, Las Vegas
Las Vegas, Nevada 89154
U.S.A.

Let $\mathbf{Z}_g$ denote the ring of g-adic integers and $\mathbf{N}$ be the set of positive integers. Throughout this paper, we consider sequences of g-adic integers $\alpha = \{a_n\}_{n=1,2,\cdots}$ and $\beta = \{b_n\}_{n=1,2,\cdots}$.

For a g-adic integer $j \in \mathbf{Z}_g$ and a positive integer $N \in \mathbf{N}$, $A_N(U_k(j) ; \alpha)$ denotes the number of terms $1 \leqq n \leqq N$ satisfying $a_n \in U_k(j)$, where

$$U_k(j) = \{x ; x \in \mathbf{Z}_g, |x-j|_g \leqq g^{-k}\} , \tag{1}$$

where $|\cdot|_g$ denotes the g-adic valuation and k is a fixed positive integer.

<u>Definition 1</u>. A sequence of g-adic integers $\alpha = \{a_n\}_{n=1,2,\cdots}$ is said to have $\{\phi_k\}$ as its asymptotic k-distribution function, if the limit

$$\lim_{N\to\infty} \frac{A_N(U_k(j) ; \alpha)}{N} = \phi_k(j) \tag{2}$$

exists for any g-adic integer $j \in \mathbf{Z}_g$.

For abbreviation, we say that this sequence $\alpha = \{a_n\}_{n=1,2,\cdots}$ has a. k-d. f. in $\mathbf{Z}_g$. We also denote $\phi_k(j)$: the limit of (2) by $\| A(\alpha \in U_k(j)) \|$ for short.

It is observed here that it is sufficient to restrict any g-adic integer j to range over the set

$$G_k = \{0, 1, 2, \cdots, g^k - 1\}, \tag{3}$$

since

$$\mathbf{Z}_g = \bigcup_{j=0}^{g^k-1} U_k(j). \tag{4}$$

<u>Definition 2</u>. Suppose that a sequence of g-adic integers $\alpha = \{a_n\}_{n=1,2,\cdots}$ has $\{\phi_k\}_{j\in \mathbf{Z}_g}$ as its asymptotic k-distribution function. If $\phi_k(j) = 1/g^k$ for every $j \in G_k$, then $\alpha = \{a_n\}_{n=1,2,\cdots}$ is called k-uniformly distributed in $\mathbf{Z}_g$. For abbreviation, we call $\alpha = \{a_n\}_{n=1,2,\cdots}$ k-u. d. in $\mathbf{Z}_g$.

<u>Definition 3</u>. If a sequence of g-adic integers $\alpha = \{a_n\}_{n=1,2,\cdots}$ is k-uniformly distributed in $\mathbf{Z}_g$ for each positive integer $k \in \mathbf{N}$, then the sequence $\alpha = \{a_n\}_{n=1,2,\cdots}$ is said to be uniformly distributed in $\mathbf{Z}_g$.

For abbreviation, we say that $\alpha = \{a_n\}_{n=1,2,\cdots}$ is u. d. in $\mathbf{Z}_g$.

For the detail see [6].

Now we consider simultaneously another sequence of g-adic integers $\beta = \{b_n\}_{n=1,2,\cdots}$. For fixed positive integers k_1 and k_2 and g-adic integers ℓ and m, we denote the number of sufixes n with $1 \leqq n \leqq N$ such that $a_n \in U_{k_1}(\ell)$ and $b_n \in U_{k_2}(m)$ by $A_N(U_{k_1}(\ell) \,;\, \alpha, U_{k_2}(m) \,;\, \beta)$. We write

$$\begin{aligned} &\lim_{N\to\infty} \frac{1}{N} A_N(U_{k_1}(\ell) \,;\, \alpha, U_{k_2}(m) \,;\, \beta) \\ &= \| A(\alpha \in U_{k_1}(\ell), \beta \in U_{k_2}(m)) \| , \end{aligned} \tag{5}$$

whenever the limit of (5) exists. We observe that if the limit (5) exists for ℓ and m running independently through G_{k_1} and G_{k_2}, respectively, then both sequences $\alpha = \{a_n\}_{n=1,2,\cdots}$ and $\beta = \{b_n\}_{n=1,2,\cdots}$ have an a. k-d. f.

In the present paper, we develop a notion of independence which is analogous to the independence of sequences of rational integers mod m (Kuipers and Shiue [3], p.63), and of Gaussian integers, (Burke and Kuipers [1]). The notion of independence in $\mathbf{Z}_g$ was also given by Shiue [6].

Definition 4. Two sequences of g-adic integers $\alpha = \{a_n\}_{n=1,2,\cdots}$ and $\beta = \{b_n\}_{n=1,2,\cdots}$ are called (k_1, k_2)-independent in $\mathbf{Z}_g$, if and only if, for any pair of g-adic integers (ℓ, m) with $\ell \in G_{k_1}$ and $m \in G_{k_2}$, the limit

$$\| A(\alpha \in U_{k_1}(\ell),\ \beta \in U_{k_2}(m)) \|$$

exists and satisfies

$$\| A(\alpha \in U_{k_1}(\ell),\ \beta \in U_{k_2}(m)) \| = \| A(\alpha \in U_{k_1}(\ell) \| \cdot \| A(\beta \in U_{k_2}(m)) \| , \qquad (6)$$

where k_1 and k_2 are fixed positive integers.

Definition 5. We call two sequences of g-adic integers $\alpha = \{a_n\}_{n=1,2,\cdots}$ and $\beta = \{b_n\}_{n=1,2,\cdots}$ independent in $\mathbf{Z}_g$, if α and β are (k_1, k_2)-independent for all pairs of positive integers $(k_1, k_2) \in \mathbf{N} \times \mathbf{N}$.

If α and β are (k, k)-independent in $\mathbf{Z}_g$, then α and β are said to be k-independent in $\mathbf{Z}_g$.

Theorem 1. *Two sequences of* g-*adic integers* $\alpha = \{a_n\}_{n=1,2,\cdots}$ *and* $\beta = \{b_n\}_{n=1,2,\cdots}$ *are* k-*independent in* $\mathbf{Z}_g$, *if and only if, for every pair of* g-*adic integers* $(h, d) \in \mathbf{Z}_g \times \mathbf{Z}_g$, *the sequence* $h\alpha + d\beta = \{ha_n + db_n\}_{n=1,2,\cdots}$ *has an* a. k-d. f. *in* $\mathbf{Z}_g$ *given by*

$$\| A(h\alpha + d\beta \in U_k(j)) \| = \sum_{\substack{\ell, m \in G_k \\ h\ell + dm \in U_k}} \| A(\alpha \in U_k(\ell)) \| \cdot \| A(\beta \in U_k(m)) \| , \tag{7}$$

for all $j \in G_k$.

Proof. First we prove the necessity. We represent two g-adic integers h and d by

$$h = \sum_{i=0}^{\infty} h_i \cdot g^i , \quad d = \sum_{i=0}^{\infty} h_i \cdot g^i .$$

Since $a_n \in U_k(\ell)$ and $b_n \in U_k(m)$, we can write

$$a_n = \ell + \sum_{i=k}^{\infty} a_n^i \cdot g^i$$

and

$$b_n = m + \sum_{i=k}^{\infty} b_n^i \cdot g^i .$$

This implies

$$\begin{aligned} ha_n + db_n &= h(\ell + \sum_{i=k}^{\infty} a_n^i \cdot g^i) + d(m + \sum_{i=k}^{\infty} b_n^i \cdot g^i) \\ &= h\ell + dm + h \cdot \sum_{i=k}^{\infty} a_n^i \cdot g^i + d \cdot \sum_{i=k}^{\infty} b_n^i \cdot g^i . \end{aligned}$$

Now, if $h\ell + dm \in U_k(j)$, i.e.

$$h\ell + dm = j + \sum_{i=k}^{\infty} c_i \cdot g^i ,$$

then

$$ha_n + db_n = j + \sum_{i=k}^{\infty} c_i \cdot g^i + h \cdot \sum_{i=k}^{\infty} a_n^i \cdot g^i + d \cdot \sum_{i=k}^{\infty} b_n^i \cdot g^i .$$

This implies that $ha_n + db_n \in U_k(j)$.

Hence

$$A_N(U_k(j)\ ;\ h\alpha + d\beta)$$
$$= \sum_{\substack{\ell, m \in G_k \\ h\ell + dm \in U_k(j)}} A_N(U_k(\ell)\ ;\ \alpha,\ U_k(m)\ ;\ \beta)\ .$$

That is

$$\| A(h\alpha + d\beta \in U_k(j)) \|$$
$$= \sum_{\substack{\ell, m \in G_k \\ h\ell + dm \in U_k(j)}} \| A(\alpha \in U_k(\ell)\ ;\ \beta \in U_k(m)) \|$$
$$= \sum_{\substack{\ell, m \in G_k \\ h\ell + dm \in U_k(j)}} \| A(\alpha \in U_k(\ell)) \| \cdot \| A(\beta \in U_k(m)) \|\ ,$$

since $\alpha = \{a_n\}_{n=1,2,\cdots}$ and $\beta = \{b_n\}_{n=1,2,\cdots}$ are k-independent in $\mathbf{Z}_g$. Thus the necessity part of the theorem is established.

Secondly, we proceed to the sufficiency. Let us consider the sum, for two g-adic integers $p,\ q \in G_k$,

$$\frac{1}{g^{2k}} \sum_{h,d \in G_k} \left[\exp\left\{ -\frac{2\pi i}{g^k} (hp + dq) \right\} \times \exp\left\{ \frac{2\pi i}{g^k} \varphi_k(hx + dy) \right\} \right] , \quad (8)$$

where $\varphi_k(a) = \sum_{i=-\infty}^{k-1} a_i \cdot g^i$ for a g-adic number $a = \sum_{i=-\infty}^{\infty} a_i \cdot g^i$, and $x,\ y$ are two g-adic integers.

In order to calculate the sum (8), we consider four cases according as x is contained in $U_k(p)$ or not and y is contained in $U_k(q)$ or not.

<u>Case I</u>: $x \in U_k(p)$ and $y \in U_k(q)$.

In this case, we can write

$$x = p + \sum_{i=k}^{\infty} x_i \cdot g^i \quad \text{and} \quad y = q + \sum_{i=k}^{\infty} y_i \cdot g^i\ .$$

Then

$$hx + dy = hp + dq + h\cdot \sum_{i=k}^{\infty} x_i\cdot g^i + d\cdot \sum_{i=k}^{\infty} y_i\cdot g^i ,$$

that is,

$$\varphi_k(hx + dy) = \varphi_k(hp + dq) .$$

Now put

$$hp + dq = j + \sum_{i=k}^{\infty} c_i\cdot g^i .$$

Then

$$\varphi_k(hx + dy) = \varphi_k(hp + dq) = j ,$$

and the sum (8) is identical with

$$\frac{1}{g^{2k}} \sum_{h,d\in G_k} \exp\left\{ \frac{2\pi i}{g^k} (-g^k\cdot \sum_{i=0}^{s-k} c_{i+k}\cdot g^i)\right\} = 1 . \qquad (9)$$

Case II: $x\in U_k(p)$ and $y\notin U_k(q)$.

Let us write

$$x = x' + \sum_{i=k}^{\infty} x_i\cdot g^i \quad \text{and} \quad y = y' + \sum_{i=k}^{\infty} y_i\cdot g^i ,$$

then

$$hx' + dy' = \varphi_k(hx' + dy') + \sum_{i=k}^{s} c_i\cdot g^i ,$$

$$= \varphi_k(hx' + dy') + g^k\cdot \sum_{i=0}^{s-k} c_{i+k}\cdot g^i .$$

In this case, the sum (8) will be reduced to zero. Since

$$(8) = \frac{1}{g^{2k}} \sum_{h,d\in G_k} \exp\left\{ \frac{2\pi i}{g^k} (hx' + dy' - hp - dq - g^k\cdot \sum_{i=0}^{s-k} c_{i+k}\cdot g^i)\right\}$$

$$= \frac{1}{g^{2k}} \sum_{h,d\in G_k} \exp\left\{ \frac{2\pi i}{g^k} \{h(x' - p) + d(y' - q)\}\right\} .$$

$$(10) = \sum_{d\in G_k}\left(\sum_{h\in G_k}\exp\left[\frac{2\pi i}{g^k}\{h(x'-p)+d(y'-q)\}\right]\right).$$

We assume that $x\in U_k(p)$ and $y\notin U_k(q)$, then $x'=p$ and $d(y'-q)\not\equiv 0 \pmod{g^k}$. Thus the sum (10) vanishes, i.e. the sum (8) is equal to zero.

<u>Case III</u>: $x\notin U_k(p)$ and $y\in U_k(q)$.
Similarly to the Case II, the sum (10) is easily seen to be zero.

<u>Case IV</u>: $x\notin U_k(p)$ and $y\notin U_k(q)$.
Combining the above Cases II and III, we observe that the sum (10), consequently the sum (8) is equal to zero.

Now we count the number of terms $a_n\in U_k(p)$ and $b_n\in U_k(q)$ with $1\leqq n\leqq N$ by using the sum (8). Indeed,

$$\begin{aligned}
&A_N(U_k(p)\ ;\ \alpha,\ U_k(q),\ \beta)\\
&= \frac{1}{g^{2k}}\cdot\sum_{h,d\in G_k}\exp\left\{-\frac{2\pi i}{g^k}(hp+dq)\right\}\\
&\qquad\times\left[\sum_{n=1}^{N}\left(\exp\left\{\frac{2\pi i}{g^k}\varphi_k(ha_n+db_n)\right\}\right]\right)\\
&= \frac{1}{g^{2k}}\cdot\sum_{h,d\in G_k}\left(\exp\left\{-\frac{2\pi i}{g^k}(hp+dq)\right\}\right.\\
&\qquad\left.\times\left\{\sum_{j\in G_k}\left[A_N(U_k(j)\ ;\ h\alpha+d\beta)\cdot\exp\left\{\frac{2\pi i}{g^k}j\right\}\right]\right\}\right).
\end{aligned}$$

We divide the above formula by N and obtain, by letting $N\to\infty$,

$$\begin{aligned}
&\|\ A(\alpha\in U_k(p),\ \beta\in U_k(q))\ \| \qquad\qquad (11)\\
&= \frac{1}{g^{2k}}\cdot\sum_{h,d\in G_k}\left(\exp\left\{-\frac{2\pi i}{g^k}(hp+dq)\right\}\right.\\
&\qquad\left.\times\left\{\sum_{j\in G_k}\left[\|\ A(h\alpha+d\beta\in U_k(j))\ \|\cdot\exp\left\{\frac{2\pi i}{g^k}j\right\}\right]\right\}\right)
\end{aligned}$$

$$= \frac{1}{g^{2k}} \cdot \sum_{h,d \in G_k} \left(\exp\left\{ -\frac{2\pi i}{g^k} (hp + dq) \right\} \right.$$

$$\times \sum_{j \in G_k} \left[\exp\left\{ \frac{2\pi i}{g^k} j \right\} \right.$$

$$\left. \left. \times \left\{ \sum_{\ell, m \in G_k} \| A(\alpha \in U_k(\ell)) \| \cdot \| A(\beta \in U_k(m)) \| \right\} \right] \right)$$

$$= \frac{1}{g^{2k}} \cdot \sum_{\ell, m \in G_k} \left(\| A(\alpha \in U_k(\ell)) \| \cdot \| A(\beta \in U_k(m)) \| \right.$$

$$\left. \times \sum_{j \in G_k} \left[\exp\left\{ \frac{2\pi i}{g^k} j \right\} \times \sum_{\substack{h,d \in G_k \\ h\ell + dm \in U_k(j)}} \exp\left\{ -\frac{2\pi i}{g^k} (hp + dq) \right\} \right] \right) ,$$

where we suppose (7) and change the order of summations.

We consider the sum $\sum_{j \in G_k} [\quad]/g^{2k}$ in the above formula:

$$\frac{1}{g^{2k}} \cdot \sum_{j \in G_k} \left[\exp\left\{ \frac{2\pi i}{g^k} j \right\} \times \sum_{\substack{h,d \in G_k \\ h\ell + dm \in U_k(j)}} \exp\left\{ -\frac{2\pi i}{g^k} (hp + dq) \right\} \right] \quad (12)$$

$$= \frac{1}{g^{2k}} \cdot \sum_{j \in G_k} \left[\sum_{\substack{h,d \in G_k \\ h\ell + dm \in U_k(j)}} \exp\left\{ \frac{2\pi i}{g^k} (j - hp - dq) \right\} \right]$$

$$= \frac{1}{g^{2k}} \cdot \sum_{h,d \in G_k} \left[\sum_{j \in G_k} \exp\left\{ \frac{2\pi i}{g^k} (j - hp - dq) \right\} \right] ,$$

by using the results for the sum (8).

Now we distinguish two cases.

Case (i): $\ell = p$ and $m = q$.

Case (ii): otherwise.

We treat the Case (i) first. Let us assume that four g-adic integers h, d, ℓ and m are given by

$$h = \sum_{i=0}^{k-1} h_i \cdot g^i , \qquad d = \sum_{i=0}^{k-1} d_i \cdot g^i ,$$

$$\ell = \sum_{i=0}^{k-1} \ell_i \cdot g^i , \qquad m = \sum_{i=0}^{k-1} m_i \cdot g^i .$$

Then, in this case

$$j - hp - dq = j - h\ell - dm .$$

But $h\ell + dm \in U_k(j)$,

$$h\ell + dm = j + \sum_{i=k}^{s} c_i \cdot g^i .$$

Thus

$$j - hp - dq = -\sum_{i=k}^{s} c_i \cdot g^i = -g^k \sum_{i=0}^{s-k} c_{i+k} \cdot g^i ,$$

that is,

$$\exp\left\{\frac{2\pi i}{g^k} (j - hp - dq)\right\} = \exp\left\{-\frac{2\pi i}{g^k} g^k \sum_{i=0}^{s-k} c_{i+k} \cdot g^i\right\} = 1 ,$$

which implies that the sum (12) is equal to 1.

Case (ii): In this case, the sum

$$\sum_{j \in G_k} \exp\left\{\frac{2\pi i}{g^k} (j - hp - dq)\right\} \tag{13}$$

is equal to zero.

Let us go back to the sum (11) and we deduce that

$$\begin{aligned} &\| A(\alpha \in U_k(p), \beta \in U_k(q)) \| \\ &= \| A(\alpha \in U_k(\ell)) \| \cdot \| A(\beta \in U_k(q)) \| , \end{aligned} \tag{14}$$

for every pair of g-adic integers $(p, q) \in \mathbf{Z}_g \times \mathbf{Z}_g$, which completes the proof.

< Q.E.D. >

Definition 6. For a fixed pair of positive integers (k_1, k_2), we call a function f from $\mathbf{Z}_g \times \mathbf{Z}_g$ to $\mathbf{Z}_g$ (k_1, k_2) congruence

preserving function if

$$f(i_1, i_2) = f(j_1, j_2) ,$$

whenever $j_1 \in U_{k_1}(i_1)$ and $j_2 \in U_{k_2}(i_2)$.

Remark. Suppose that two sequences of g-adic integers $\alpha = \{a_n\}_{n=1,2,\cdots}$ and $\beta = \{b_n\}_{n=1,2,\cdots}$ are (k_1, k_2)-independent in $\mathbf{Z}_g$, then the sequence $f(\alpha, \beta) = \{f(a_n, b_n)\}_{n=1,2,\cdots}$ has an a. k-d. f. in $\mathbf{Z}_g$ for any (k_1, k_2) congruence preserving function f.

Indeed,

$$A_N(U_k(j) ; f(\alpha, \beta)) = \sum_{\substack{\ell \in G_{k_1} \\ m \in G_{k_2} \\ f(\ell,m) \in U_k(j)}} A_N(U_{k_1}(\ell) ; \alpha, U_{k_2}(m) ; \beta) .$$

From the assumption of (k_1, k_2)-independence, we deduce that

$$\| A(f(\alpha, \beta) \in U_k(j)) \| = \sum_{\substack{\ell \in G_{k_1} \\ m \in G_{k_2} \\ f(\ell,m) \in U_k(j)}} \| A(\alpha \in U_{k_1}(\ell)) \| \cdot \| A(\beta \in U_{k_2}(m)) \| ,$$

which indicates the existence of the a. k-d. f. of $f(\alpha, \beta)$ in Z_g .

Theorem 2. *Let us assume that two sequences of* g-*adic integers* $\alpha = \{a_n\}_{n=1,2,\cdots}$ *and* $\beta = \{b_n\}_{n=1,2,\cdots}$ *are* (k_1, k_2)-*independent in* $\mathbf{Z}_g$. *Then, for any pair of* g-*adic integers* $(h, d) \in \mathbf{Z}_g \times \mathbf{Z}_g$, $h\alpha = \{ha_n\}_{n=1,2,\cdots}$ *and* $d\beta = \{db_n\}_{n=1,2,\cdots}$ *are* (k_1, k_2)-*independent in* $\mathbf{Z}_g$.

Proof. Without loss of generality, we may assume that $\ell \in G_{k_1}$ and $m \in G_{k_2}$ for any pair of positive integers $(k_1, k_2) \in \mathbf{N} \times \mathbf{N}$.

We consider two sets of g-adic integers X and Y defined by

$$X = \{x \; ; \; x \in G_{k_1} \quad \text{and} \quad hx \in U_{k_1}(\ell)\} \; ,$$

$$Y = \{y \; ; \; y \in G_{k_2} \quad \text{and} \quad dy \in U_{k_2}(\ell)\} \; .$$

We represent g-adic integers a_n and b_n by

$$a_n = \sum_{i=0}^{\infty} a_n^i \cdot g^i \; ,$$

$$b_n = \sum_{i=0}^{\infty} b_n^i \cdot g^i \; .$$

Now we distinguish two cases according as two sets X and Y are non-empty or not.

Case I: X and Y are both non-empty.
Since for all $x \in X$, $hx \in U_{k_1}(\ell)$, then

$$hx = \ell + \sum_{i=k}^{\infty} c_i \cdot g^i \; .$$

If $a_n \in U_{k_1}(x)$, then

$$a_n = x + \sum_{i=k}^{\infty} a_n \cdot g^i \; ,$$

so that

$$ha_n = hx + h \cdot \sum_{i=k}^{\infty} a_n^i \cdot g^i$$

$$= \ell + h \cdot \sum_{i=k}^{\infty} a_n^i \cdot g^i + \sum_{i=0}^{\infty} c_i \cdot g^i \; .$$

This implies that

$$\varphi_k(ha_n) = \ell \; ,$$

that is

$$ha_n \in U_{k_1}(\ell) \; .$$

Similarly for $y \in Y$ and $b_n \in U_{k_2}(y)$, we have $db_n \in U_{k_2}(m)$. Thus we obtain

$$A_N(U_{k_1}(\ell) ; h\alpha, U_{k_2}(m) ; d\beta) = \sum_{\substack{x \in X \\ y \in Y}} A_N(U_{k_1}(x) ; \alpha, U_{k_2}(y) ; \beta) .$$

Dividing the above formula by N and then letting $N \to \infty$, we get

$$\begin{aligned} &\| A(h\alpha \in U_{k_1}(\ell), d\beta \in U_{k_2}(m)) \| \\ &= \sum_{\substack{x \in X \\ y \in Y}} \| A(\alpha \in U_{k_1}(x), \beta \in U_{k_2}(y)) \| \\ &= \left(\sum_{x \in X} \| A(\alpha \in U_{k_1}(x)) \| \right) \left(\sum_{y \in Y} \| A(\beta \in U_{k_2}(y)) \| \right) \\ &= \| A(h\alpha \in U_{k_1}(\ell)) \| \cdot \| A(d\beta \in U_{k_2}(m)) \| , \end{aligned}$$

by the (k_1, k_2)-independence of α and β. This indicates that the two sequences of g-adic integers $h\alpha$ and $d\beta$ are (k_1, k_2)-independent in $\mathbf{Z}_g$.

Case II: X or Y is empty.
In this case,

$$\| A(h\alpha \in U_{k_1}(\ell), d\beta \in U_{k_2}(m)) \| = 0 .$$

On the other hand, one at least of

$$\| A(h\alpha \in U_{k_1}(\ell)) \| , \| A(d\beta \in U_{k_2}(m)) \|$$

is equal to zero. Consequently

$$\begin{aligned} &\| A(h\alpha \in U_{k_1}(\ell), d\beta \in U_{k_2}(m)) \| \\ &= \| A(h\alpha \in U_{k_1}(\ell)) \| \cdot \| A(d\beta \in U_{k_2}(m)) \| = 0 , \end{aligned}$$

which means the (k_1, k_2)-independence of $h\alpha$ and $d\beta$.

< Q.E.D. >

Theorem 3. *Fix a pair of positive integers* $(k_1, k_2) \in \mathbf{N} \times \mathbf{N}$ *with* $k_1 \leq k_2$. *Suppose that a sequence of* g-*adic integers* $\alpha = \{a_n\}_{n=1,2,\cdots}$ *has* $\{\phi_{k_1}\}$ *as its* a. k_1-d. f. *in* $\mathbf{Z}_g$. *Then* $\alpha = \{a_n\}_{n=1,2,\cdots}$

and α *are* (k_1, k_2)*-independent in* $\mathbf{Z}_g$, *if and only if* $\phi_{k_1}(j) = 1$ *for some* $j \in G_{k_1}$.

Proof. We prove first the necessity part by contradiction. For every g-adic integer $\ell \in G_{k_1}$, we assume that $0 < \phi_{k_2}(\ell) < 1$. We write

$$a_n = \sum_{i=0}^{\infty} a_n \cdot g^i .$$

Let us calculate $A_N(U_{k_2}(\ell) ; \alpha)$. If $a_n \in U_{k_2}(\ell)$, then a_n can be written by

$$a_n = \ell + \sum_{i=k_2}^{\infty} a_n^i \cdot g^i \quad \text{with} \quad \ell \in G_{k_1} .$$

By setting $a_n^i = 0$ for i with $k_1 \leq i \leq k_2-1$, we can rewrite a_n by

$$a_n = \ell + \sum_{i=k_1}^{\infty} a_n^i \cdot g^i ,$$

which implies that $a_n \in U_{k_1}(\ell)$.

Thus

$$A_N(U_{k_2}(\ell) ; \alpha) = A_N(U_{k_1}(\ell) ; \alpha, U_{k_2}(\ell) ; \alpha) .$$

We divide the above formula by N and let N tend to infinity to get

$$\| A(\alpha \in U_{k_2}(\ell)) \| = \| A(\alpha \in U_{k_1}(\ell), \alpha \in U_{k_2}(\ell)) \|$$

$$= \| A(\alpha \in U_{k_1}(\ell)) \| \cdot \| A(\alpha \in U_{k_2}(\ell)) \| ,$$

if $\alpha = \{a_n\}_{n=1,2,\ldots}$ and α are (k_1, k_2)-independent in $\mathbf{Z}_g$. But the above formula is identical with

$$\phi_{k_2}(\ell) = \phi_{k_1}(\ell) \cdot \phi_{k_2}(\ell)$$

with $0 < \phi_{k_2}(\ell) < 1$. This means that

$$\phi_{k_1}(\ell) = 1 \quad \text{for every} \quad \ell \in G_{k_1} ,$$

which leads to a contradiction to the fact that α has $\{\phi_{k_1}\}$ as its a. k_1-d. f. in $\mathbf{Z}_g$. Thus we finish the proof for the necessity part.

In order to show the sufficiency, we assume that $\phi_{k_2}(j) = 1$ for some $j \in G_{k_1}$. Let us take two g-adic integers $\ell \in G_{k_1}$ and $m \in G_{k_2}$.

Case (i): $\ell \neq m$.

In this case, ℓ and m cannot be simultaneously equal to j and from the fact that $\phi_{k_2}(j) = 1$ for some $j \in G_{k_1}$, we deduce that

$$\| A(\alpha \in U_{k_1}(\ell), \alpha \in U_{k_2}(m)) \| = 0$$

and at least one of $\| A(\alpha \in U_{k_1}(\ell)) \|$, $\| A(\alpha \in U_{k_2}(m)) \|$ is equal to zero. Consequently

$$\begin{aligned} &\| A(\alpha \in U_{k_1}(\ell), \alpha \in U_{k_2}(m)) \| \\ &= \| A(\alpha \in U_{k_1}(\ell)) \| \cdot \| A(\alpha \in U_{k_2}(m)) \| = 0 , \end{aligned}$$

which indicates the (k_1, k_2)-independence of α and α .

Case (ii): $\ell = m \neq j$.

By almost all the same argument as in the Case (i), we obtain that

$$\begin{aligned} &\| A(\alpha \in U_{k_1}(\ell), \alpha \in U_{k_2}(\ell)) \| \\ &= \| A(\alpha \in U_{k_1}(\ell)) \| \cdot \| A(\alpha \in U_{k_2}(\ell)) \| = 0 . \end{aligned}$$

Case (iii): $\ell = m = j$.

Since $\phi_{k_2}(j) = 1$ for some $j \in G_{k_1}$,

$$\| A(\alpha \in U_{k_1}(j)) \| = \| A(\alpha \in U_{k_2}(j)) \| = 1 .$$

Similarly we have

$$\| A(\alpha \in U_{k_1}(j), \alpha \in U_{k_2}(j)) \| = 1 ,$$

so that

$$\| A(\alpha \in U_{k_1}(j), \alpha \in U_{k_2}(j)) \|$$
$$= \| A(\alpha \in U_{k_1}(j)) \| \cdot \| A(\alpha \in U_{k_2}(j)) \| = 1 ,$$

which signifies the (k_1, k_2)-independence of $\alpha = \{a_n\}_{n=1,2,\cdots}$ and α .

All the cases having been examined, we thus complete the proof of the sufficiency part.

< Q.E.D. >

REFERENCES

[1] Burke, J. R. and Kuipers, L.: Asymptotic distribution and independence of sequences of Gaussian integers, Simon Stevin, 50, 3-21 (1976).

[2] Kuipers, L. and Niederreiter, H.: Asymptotic distribution mod m and independence of sequences of integers, I, II, Proc. Japan Acad., 50, 256-260; 261-265 (1974).

[3] Kuipers, L. and Shiue, J.-S.: Asymptotic distribution mod m of sequences of integers and the notion of independence, Atti Accad. Naz. Lincei, (8) 11, 63-90 (1972).

[4] Nathanson, M. B.: Asymptotic distribution and asymptotic independence of sequences of integers, Acta Math. Acad. Sci. Hungar., 29, 207-218 (1977).

[5] Niederreiter, H.: Independance de suites, Repartition Modulo 1, Lecture Notes in Math., 475, 120-131 (1975).

[6] Shiue, J.-S.: On a theorem of uniform distribution of g-adic integers and a notion of independence, Rend. Accad. Naz. Lincei, 50, 90-93 (1971).

Proc. Prospects
of Math. Sci.
World Sci. Pub.
173-180, 1988

A PERIODIC SOLUTION OF SOME CONGRUENCES AND THE RANK OF HASSE-WITT MATRIX

TORU NAKAHARA

Department of Mathematics
Faculty of Science and Engineering
Saga University
Saga 840
Japan

ABSTRACT

The aim of this paper is to give a generalization of results in a paper [3] concerning a periodic solution of some simultaneous congruences. As an application of our theorems we shall characterize some algebraic function fields whose Hasse-Witt matrices do not have the full rank but have a positive rank.

1. INTRODUCTION

Let f and p be an odd prime power and an odd prime respectively and relatively prime.

First we consider the following r simultaneous congruences:

$$pj_k \equiv j_{k+1} \bmod f, \tag{1}_r$$

where $\{j_k\}_{k \bmod r}$ $\{1, \cdots, (f-1)/2\}$ and $r = \mathrm{Ind}_f\, p$.

When there exists such a solution $\{j_k\}$ of 'positive' absolutely least residue modulo f, we say that the congruences $(1)_r$ have a periodic solution and r is the length of the period.

In Section 2 we shall solve the simultaneous congruences $(1)_r$ for the length r equal to 3 and for prime numbers f and p greater than 2.

In Section 3, by applying periodic solutions of the congruences $(1)_r$, we shall give an estimate for the rank of the Hasse-Witt matrices of some algebraic function fields.

2. A PERIODIC SOLUTION OF SIMULTANEOUS CONGRUENCES

We denote by $R^\times$ the group of the reduced residue classes modulo f and identify each class with its representative. For any $a, b \in R^\times$ we put $a \equiv w^\alpha$ and $b \equiv w^\beta$ mod f, where w is a primitive root of modulo f. Now if $\alpha \equiv \beta \mod \frac{\varphi(f)}{r}$, we say that a and b are equivalent. Here φ denotes the Euler function. Then we can define an equivalence relation in $R^\times$.

For any $a \in R^\times$ we have an equivalence class:

$$C_a = \{a, pa, \cdots, p^{r-1}a\}.$$

Then it holds

$$R^\times = \bigcup_{a \in R} C_a \quad \text{(disjoint union).} \tag{$(2)_r$}$$

When the length r of $(1)_r$ is even, any class C_a in $(2)_r$ is not included in $\{1, \cdots, (f-1)/2\}$. Hence r is odd.

Theorem 1. *Let* f *be an odd prime power whose prime factor is congruent to* 1 *modulo* 6. *When the length* r *of the period in* $(1)_r$ *is equal to* 3, *we can get a periodic solution of* $(1)_3$ *for all prime powers* f *up to* f = 7.

Proof. Let p be a prime congruent to $w^{\varphi(f)/3}$ or $w^{2\varphi(f)/3}$ modulo f. Then there exist uniquely $a, b \in [1, g]$, $g = (f-1)/2$ such that $p \equiv \pm a$ and $p^2 \equiv \pm b$ mod f. It is enough for us to consider the case where $p \equiv a$ and $p^2 \equiv -b$ mod f, namely

$$C_1 = \{1, a, -b\} \text{ and } 1 + a - b = 0. \tag{3}$$

Because the case where $p \equiv -a$ and $p^2 \equiv -b$ mod f is impossible and in the case where $p \equiv a$ and $p^2 \equiv b$ mod f, we may consider to replace p by p^2. For a class $C_j = \{j, pj, p^2j\}$ in the partition $(2)_3$ and an integer c, cC_j means the class $\{cj, cpj, cp^2j\}$ and

$C_i + C_j$ signifies the class $\{i + j, p(i + j), p^2(1 + j)\}$.

Now we select a class $C_a + C_b = C_a + (-C_{-b})$ $= \{a + b, -b - 1, 1 - a\}$. Hence we notice $-1 \equiv w^{\varphi(f)}$ mod f.

i) For $a + b > g$ and $b < g$, we have a solution $-(C_a + C_b)$ $= \{-(a + b), b + 1, -1 + a\} = \{f - (a + b), b + 1, -1 + a\}$ of $(1)_3$.

ii) For $b = g$ it holds $a = g - 1$. Since $-b \equiv a^2$, $0 \equiv g^2 - 3g \bmod f$ holds. Then $0 \equiv 4(g^2 - 3g) \equiv 7 \bmod f$. This implies $f = 7$, in which there exists no solution of $(1)_3$.

iii) For $a + b < g$ we consider the case $\text{iii})_1$ where f is a prime and the case $\text{iii})_2$ where f is a prime power.

$\text{iii})_1$ In this case we choose the smallest integer d such that $d(a + b) > f/2$, and then we obtain a solution $-d(C_a + C_b)$ $= \{f - d(a + b), d(b + 1), d(-1 + a)\}$. The condition $d(b + 1) < f/2$ is valid for $b \geqq 5$. In fact $2d(b + 1) \leqq 2(d - 1)(a + b)$ holds if and only if $2 + (3/(b - 2)) \leqq d$ is satisfied. For $b \geqq 5$ the value $d = 3$ satisfies the latter inequality. On the other hand the situation (3) implies $b \neq 1, 2$. For the case of $b = 3$, the congruence $-b \equiv a^2$ is reduced to $-3 \equiv 4 \bmod f$, namely $f = 7$, which however is an exceptional case. For the case of $b = 4$, we get $f = 13$. For the case of $d = 2$ if $d(b + 1) > f/2$, it holds that $4b + 4 > f > 4b - 2$. The last inequality follows from the definition of d. Then for the three cases for which $f = 4b + 3$, $4b + 1$ and $4b - 1$, it follows that $f = 37$, 21 and 13, respectively. However these three values of f are excluded, since $27 \not\equiv 3 \bmod 4$, $21 \neq$ a prime power and we have $a = 3$, $b = 4$ for $f = 13$, which is a case of i) (cf. Remark 1).

$\text{iii})_2$ In this case we select the same number d as in $\text{iii})_1$. It is enough for us to consider the case for which d and f are not relatively prime. From the assumption, we write, $f = q^e$ with a prime q and $q \geqq 7$ hold. Put $d' = d + 1$, then d' and f are relatively prime. The condition $d'(b + 1) < f/2$ is valid for $b \geqq 4$. Indeed $2d'(b + 1) \leqq 2(d' - 2)(a + b)$ holds if and only if $4 + (6/(b - 2)) \leqq d'$ is satisfied. For $b > 3$ the value $d' = 7$

satisfies the latter inequality. The case of $b = 3$ i.e. $f = 7$ is an exceptional case.

By i), ii) and iii) we have finished the proof of Theorem 1.

< Q.E.D. >

Remark 1. All the periodic solutions C_j of $(1)_3$ for the next three prime powers f are as follows:

$f = 13$	$C_2 = \{2, 6, 5\}$		$p \equiv 3, -4 \bmod 13$,
$f = 7^2$	$C_4 = \{4, 22, 23\}$	$C_{12} = \{12, 17, 20\}$	$p \equiv -19, 18 \bmod 49$,
$f = 13^2$	$C_{64} = \{64, 49, 56\}$	$C_{80} = \{80, 19, 70\}$	$C_{79} = \{79, 42, 48\}$
	$C_{57} = \{57, 41, 71\}$	$C_{63} = \{63, 72, 34\}$	$p \equiv -23, 22 \bmod 169$.

Remark 2. The case for which a prime $f \equiv 19 \bmod 24$ in [3] is the prototype of Theorem 1. We can extend this case to the case where f is a prime power f by making a slight modification of the proof in [3].

Theorem 2. $f = 1 + a + \cdots + a^{r-1}$ *be a prime power for* $a > 2$ *and an odd number* $r > 1$. *Then* a *is not a square number,* r *is an odd prime, and we get a periodic solution of* $(1)_r$.

Proof. From the assumption, $f = (a^r - 1)/(a - 1) = q^e$ holds, where q is an odd prime. If an odd number $r = st > 1$ is not a prime, then $f = (a^s - 1)(a^{s(t-1)} + a^{s(t-2)} + \cdots + a^s + 1)/(a - 1)$, where s and t are larger than 1. By $a^s - 1 \equiv 0 \bmod q$, it holds that $a^{s(t-1)} + a^{s(t-2)} + \cdots + a^s + 1 \equiv 1 + 1 + \cdots + 1 + 1 \equiv t \equiv 0 \bmod q$, from which we deduce $r = q^h$, with $h > 1$. Then $a^q - 1 \equiv 0 \bmod q$ holds. Thus we get $a \equiv 1 \bmod q$ since $a^{\varphi(q)} \equiv 1 \bmod q$. For $a = 1 + q^u m$, with $(q,m) = 1$, we can see that $a^r - 1$ is exactly divisible by q^{h+u}, but $f(a-1)$ is divisible by q^{e+u}, which is a contradiction. On the other hand if a is equal to a square number b^2, then $f = \frac{(b^r - 1)(b^r + 1)}{(b - 1)(b + 1)}$. Hence $b^r - 1 \equiv b^r + 1 \equiv 0 \bmod q$, i.e. $2 \equiv 0 \bmod q$, which is impossible.

Let p be a prime congruent to a. C_1 denotes an equivalence class $\{1, a, a^2, \cdots, a^{r-1}\}$ of $(2)_r$.

i) The case where $a > 1$ is odd. Then for $c = (a + 1)/2$, c and f are relatively prime. Because if there exists a common prime factor q of c and f, then $a \equiv -1 \bmod q$, hence $f \equiv 1 - 1 + 1 - \cdots + 1 \equiv 1 \bmod q$ follows, which is a contradiction. Thus $cC_1 \in R^{\times}$. Moreover this class gives a solution:

$$\{c, ca, \cdots, ca^{r-2}, ca^{r-1} - (c - 1)f\},$$

of $(1)_r$ with the length r of the period, since all the terms are included in $[1, g]$, where $g = (f - 1)/2$.

ii) The case where $a > 4$ is even. Then for $c' = (3a + 2)/2$ and $c'' = (5a + 4)/2$, c' or c'' and f are relatively prime. Because if c', c'' and f have a common prime factor q, then $a \equiv -1 \bmod q$, hence $f \equiv 1 \bmod q$, which is impossible.

In the former case the class $c'C_1$ gives a solution:

$$\{c', c'a, \cdots, c'a^{r-2} - f, c'a^{r-1} - (c' - 2)f\},$$

where all the terms are included in $[1, g]$ for $a > 2$.

In the latter case the class $c''C_1$ gives a solution:

$$\{c'', c''a, \cdots, c''a^{r-2} - 2f, c''a^{r-1} - (c'' - 3)f\},$$

where all the terms are included in $[1, g]$ for $a > 4$.

Therefore we have furnished the proof of Theorem 2.

< Q.E.D. >

3. THE RANK OF HASSE-WITT MATRIX

In this Section we shall give an estimate for the rank of Hasse-Witt matrix of some algebraic function fields. Theorems 3 and 4 shall be obtained immediately by an application of our results in the previous section [cf. 2].

Let K be a finite field of characteristic $p > 2$ and $A_f = K(x, y)$ be an algebraic function field over K defined by $y^2 = x^f + a$ $(a \in K, a \neq 0)$, where $(p, f) = 1$ and $f = 2g + 1$ is a prime power.

Let $\omega_1 = dx/y$, $\omega_2 = xdx/y$, $\cdots$, $\omega_g = x^{g-1}dx/y$ be a basis of the K-module of holomorphic differentials in A_f. Then the Hasse-Witt matrix of A_f is defined by the representation matrix over K of the Cartier operator C with respect to a basis $\{\omega_j\}_{1 \leq j \leq g}$, [1], [6], [7]:

$$^t(C(\omega_1), \cdots, C(\omega_g)) = M^t(\omega_1, \cdots, \omega_g).$$

For any differential $\omega = (a_0^p + a_1^p + \cdots + a_{p-1}^p x^{p-1})dx \quad (a_i \in A_f)$, the operator C is defined by

$$C(\omega) = a_{p-1}dx.$$

Then we have

$$\omega_j = x^{j-1}dx/y = \sum_{k=0}^{\ell} \binom{\ell}{k} a^{\ell-k} x^{j+fk-1} dx/y^p, \quad p = 2\ell + 1,$$

and

$$C(x^{i+fk-1}dx) = \begin{cases} x^{((i+fk)/p) - 1}dx, & i + f \equiv 0 \bmod p \\ 0 & , \text{otherwise} \end{cases}.$$

Recently T. Kodama and T. Washio obtained the next three Lemmas [7].

Lemma 1. (i) *The Hasse-Witt matrix* M *has at most one non-zero element in each row and in each column*

(ii) rank $M = \#\{(i, k) \mid i + fk \equiv 0 \bmod p, (i, k) \in [1, g] \times [0, \ell]\}$.

(iii) rank $M = \#\{(i, j) \mid i \equiv pj \bmod f, (i, j) \in [1, g] \times [1, g]\}$.

Lemma 2. *The rank of* M *is equal to* g *if and only if* $p \equiv 0 \bmod f$. *In this case*

$$M = \begin{pmatrix} a_{11} & & \\ & \ddots & \\ & & a_{gg} \end{pmatrix},$$

where $a_{jj} = \binom{\ell}{k} a^{(\ell-k)/p}$ *and* $k = (p-1)j/f$.

In this case it is called that the algebraic function field A_f is normal. When A_f is not normal, we say that A_f is singular.

Lemma 3. *The rank of* M *is zero if and only if* $p \equiv -1 \bmod f$.

It is known that the algebraic function field with M of rank zero is supersingular [4], [5], [8].

We shall construct some type of algebraic function fields which are singular but not supersingular. Finding a periodic solution $\{j_k\}_{k \bmod r}$, $1 \leqq j_k \leqq g$, $j_i \not\equiv j_k \bmod f$ $(i \not\equiv k \bmod r)$ of the congruences

$$pj_k \equiv j_{k+1} \bmod f, \qquad (1)_r$$

we can pursue our purpose.

Proposition 1. *If the simultaneous congruences*

$$pj_1 \equiv j_2,\ pj_2 \equiv j_3,\ \cdots,\ pj_r \equiv j_1 \bmod f$$
$$1 \leqq j_k \leqq g \quad (k \bmod r)$$

have a periodic solution $\{j_k\}_{k \bmod r}$, *where* $r = \mathrm{Ind}_f\, p$, *then the rank of arbitrary power of the Hasse-Witt matrix* M *is not smaller than the length* r *of the period.*

Proof. By Lemma 1 and the assumption, in any j_k-th row of M only the (j_k, j_{k-1})-component has non-zero element. Then from a solution $C_{j_1} = \{j_k\}$ we obtain the diagonal matrix M^r whose rank is at least r.

Remark 3. From the above proof the rank of M^r is a multiple of r. On the other hand when we have no solution of $(1)_r$, the rank of M^r is equal to 0.

Theorem 3. *There exist infinitely many algebraic function fields* A_f *whose Hasse-Witt matrices are singular but of rank at least* 3 *for all prime powers* f *up to* $f = 7^e$ *with* $e = 1$, *where the prime factor of* f *is congruent to* 1 *modulo* 6.

Proof. This theorem follows from Theorem 1, Proposition 1 and Lemma 2.

Remark 4. In the exceptional case we have no solution of $(1)_3$ for $g = 3$. Thus the algebraic function field A_7 is supersingular from Remark 3 [8].

Combining Theorem 2, Proposition 1 and Lemma 2 we obtain the next theorem.

Theorem 4. *Let* $f = 1 + a + \cdots + a^{r-1}$ *be a prime power with* $a > 2$ *not being a square number and* r *be an odd prime number. Then there exists an algebraic function field* A_f *whose Hasse-Witt matrix is singular but of rank at least* r.

Remark 5. From Theorems 3, 4 and Proposition 1 it is shown that there exist infinitely many algebraic function fields which are singular but not supersingular.

Acknowledgement. The author should express his hearty thanks to referees for their constructive criticism including a part of the proof of Theorem 2.

REFERENCES

[1] Kodama, T.: On the rank of the Hasse-Witt matrix, Proc. Japan Acad. Ser. A Math. Sci., 60, 165-167 (1984).

[2] Miller, L.: Curves with invertible Hasse-Witt-matrix, Math. Ann., 197, 123-127 (1972).

[3] Nakahara, T.: On a periodic solution of some congruences, Rep. Fac. Sci. Engrg. Saga Univ. Math., 14, 1-5 (1986).

[4] Rosen, M.: The asymptotic behavior of the class group of a function field over a finite field, Arch. Math., 24, 287-296 (1974).

[5] Silverman, J.H., The Arithmetic of Elliptic Curves, Springer-Verlag, New York-Berlin-Heidelberg-Tokyo, 1986.

[6] Stichtenoth, H.: Die Hasse-Witt Invariante eines Kongruenzfunktionenkörpers, Arch. Math., 33, 357-360 (1979).

[7] Washio, T. and Kodama, T.: Hasse-Witt matrices of hyperelliptic function fields, Sci. Bull. Fac. Ed. Nagasaki Univ., 37, 9-15 (1986).

[8] Washio, T. and Kodama, T.: A note on a supersingular function field, ibid., 37, 17-21 (1986).

Proc. Prospects
of Math. Sci.
World Sci. Pub.
181-188, 1988

THE TAYLOR COEFFICIENTS OF SOME DIRICHLET SERIES

MASUMI NAKAJIMA

Department of Mathematics
Rikkyo University (St. Paul's Univ.)
Nishi-Ikebukuro, Toshima-ku, Tokyo 171
Japan

1. INTRODUCTION AND RESULTS

In 1905, the following formulas concerning the Riemann zeta-function $\zeta(s)$ are exposed by T.J. Stieltjes in "Correspondance d'Hermite et de Stieltjes", which was published after his death. The Stieltjes formulas are as follows:

Let γ_k be the k-th coefficient in the Laurent expansion of $\zeta(s)$ at $s=1$, where

$$\zeta(s) = \frac{1}{s-1} + \sum_{k=0}^{\infty} \frac{(-1)^k}{k!} \gamma_k (s-1)^k . \tag{1}$$

Then γ_k's are expressed by

$$\gamma_k = \lim_{x\to\infty} \{ \sum_{n\leqq x} \frac{(\log n)^k}{n} - \frac{(\log x)^{k+1}}{k+1} \} , \quad k=0,1,2,\cdots , \tag{2}$$

and γ_k is called the k-th generalized Euler constant. The constants γ_k's have been considered by many mathematicians, but the coefficients $D(k,s_0)$'s in the Taylor expansion of $\zeta(s)$ at $s_0 \neq 1$ are treated only by D. Mitrovic [2] for the case: $\mathrm{Re}\, s_0 > 1$, where

$$\zeta(s) = \sum_{k=0}^{\infty} \{-(1-s_0)^{-k-1} + (-1)^k (k!)^{-1} D(k,s_0)\} (s-s_0)^k , \quad s_0 \neq 1 , \tag{3}$$

and he obtained an upper bound estimate of the coefficients. The author obtained a stronger estimate than that of Mitrovič in [3] and afterwards improved this result by using the complex integral expression and the saddle point method, which shall be discussed elsewhere [4]. The Stieltjes formulas have been thought as the only case in which they can exist, since the point $s=1$ is the only pole of $\zeta(s)$. Recently the author obtained the following formulas in [3] which are similar to Stieltjes':

Theorem 1. *The k-th Taylor coefficient* $D(k,s_0)$ *in* (3) *can be expressed by*

$$D(k,s_0) = \lim_{x\to\infty} \{ \sum_{n\leq x} \frac{(\log n)^k}{n^{s_0}} - \int_1^x \frac{(\log u)^k}{u^{s_0}} du\} , \quad \textit{for} \quad \mathrm{Re}\, s_0 > 0 ,$$

$$D(k,s_0) = \lim_{x\to\infty} \{ \sum_{n\leq x} \frac{(\log n)^k}{n^{s_0}} - \int_1^x \frac{(\log u)^k}{u^{s_0}} du - \frac{(\log x)^k}{2x^{s_0}} \} ,$$

$$\textit{for} \quad -1 < \mathrm{Re}\, s_0 \leqq 0 ,$$

and

$$D(k,s_0) = \lim_{x\to\infty} \{ \sum_{n\leq x} \frac{(\log n)^k}{n^{s_0}} - \int_1^x \frac{(\log u)^k}{u^{s_0}} du - \frac{(\log x)^k}{2x^{s_0}} ,$$

$$- \sum_{m=2}^{M} (-1)^m \frac{B_m}{m!} \frac{d^{m-1}}{dx^{m-1}} (\frac{(\log x)^k}{x^{s_0}})\} ,$$

$$\textit{for} \quad -M < \mathrm{Re}\, s_0 \leqq -(M-1) ,$$

where M *is an integer greater than one and* B_m *is the m-th Bernoulli number defined by*

$$z/(e^z-1) = \sum_{n=0}^{\infty} (n!)^{-1} B_n z^n , \quad |z| < 2\pi .$$

For the proof of Theorem 1, see [3] and also [4], which is simpler than that of [3].

In this note, we will give Theorem 2 and Theorem 3 which partly include Theorem 1.

Theorem 2. *Suppose that the Dirichlet series*

$$F(s) = \sum_{n=1}^{\infty} a(n)(n+a)^{-s}, \quad 0 \leqq a < 1 ,$$

satisfies the following two conditions:

I. $F(s)$ *has an analytic continuation in* $\mathrm{Re}\, s > \gamma$, *with* $\gamma \leqq \beta \leqq \alpha$.

II. *Set* $A(y) = \sum_{n \leqq y} a(n)$. *Then* $A(y)$ *can be written by* $A(y) = B(y) + B_1(y)$, *where* $B(y)$ *is real analytic in* $y > 0$, *then* $B(y)$ *and* $B_1(y)$ *satisfy*

$$B(y) = O(y^{\alpha}) = \Omega(y^{\alpha}) ,$$

$$B_1(y) = O(y^{\beta}) , \text{ with } 0 \leqq \beta \leqq \alpha ,$$

and

$$B_1(y)/B(y) \to 0 , \text{ as } y \to \infty .$$

Let us put $f(s) = s\int_1^{\infty} (y+a)^{-s-1} B(y)dy$, *for* $\mathrm{Re}\, s > \alpha$.

Then the Taylor expansion of $F(s)$ *is given by:*

$$F(s) = \sum_{k=0}^{\infty} (k!)^{-1}\{f^{(k)}(s_0) - (-1)^{k+1}B(1)(1+a)^{k-s_0}\delta(a,k)$$

$$+ (-1)^k b(k,s_0)\}(s-s_0)^k, \quad \textit{for} \ \mathrm{Re}\, s_0 > \beta ,$$

where $\delta(a,k)$ *is* 1 *if* $a \neq 0$ *and* $\delta(0,k) = 0$ *if* $k > 0$, *and we put* $\delta(0,0) = 1$.

Further we have:

$$b(k,s_0) = \lim_{x\to\infty} \{ \sum_{n \leqq x} a(n)(\log(n+a))^k (n+a)^{-s_0}$$

$$- \int_1^{x} B^{(1)}(y)(\log(y+a))^k (y+a)^{-s_0} dy\} ,$$

$$\textit{for} \ \mathrm{Re}\, s_0 > \beta .$$

Remark. In most cases, $f(s)$ becomes an elementary function, so $f^{(k)}(s)$ can be easily calculated explicitly. See Examples.

In order to extend Theorem 2 into the case $\operatorname{Re} s_0 > \gamma$, we need some additional conditions on $B_1(y)$.

Theorem 3. *Suppose that the conditions on* $F(s)$ *in Theorem* 2 *hold. Further we suppose that*

$$B_{n+1}(y) = O(y^{\beta+n(1-\delta)}) , \quad \delta > 0 ,$$

for every positive integer n , *where*

$$B_{n+1}(y) = \int_1^y B_n(y)dy .$$

Then we have, for some positive integer N *with* $\gamma \leqq \beta - N\delta$,

$$\begin{aligned} b(k,s_0) = \lim_{x\to\infty} [& \sum_{n\leqq x} a(n)(\log(n+a))^k (n+a)^{-s_0} \\ & - \int_1^x B^{(1)}(y)(\log(y+a))^k (y+a)^{-s_0} dy \\ & - B(1)(\log(1+a))^k (1+a)^{-s_0} \delta(a,k) \\ & + \sum_{k=1}^{N} (-1)^k B_k(x)(d^{k-1}/dx^{k-1})\{(\log(x+a))^k (x+a)^{-s_0})\}] , \end{aligned}$$

for $\operatorname{Re} s_0 > \beta - N\delta$.

Proof of Theorem 2. By using the Stieltjes integral, we have for any small positive ε ,

$$\begin{aligned} F(s) &= \int_{1-\varepsilon}^{\infty} (y+a)^{-s} dA(y) , \\ &= s\int_1^{\infty} (y+a)^{-s-1} A(y)dy , \quad \text{for } \operatorname{Re} s > \alpha . \end{aligned}$$

Then we consider the sum $\sum_{n\leqq x} a(n)(n+a)^{-s}$, which is, for arbitrary positive small ε , equal to

$$A(x)(x+a)^{-s} + s\int_{1-\varepsilon}^{x} (y+a)^{-s-1}(B(y) + B_1(y))dy$$

$$= (1+a)^{-s}B(1) + \int_1^x (y+a)^{-s}B^{(1)}(y)dy + B_1(x)(x+a)^{-s}$$

$$+ s\int_1^\infty (y+a)^{-s-1}B_1(y)dy + \int_x^\infty (d/dy)\{(y+a)^{-s}\}B_1(y)dy$$

$$= (1+a)^{-s}B(1) + \int_1^x (y+a)^{-s}B^{(1)}(y)dy + B_1(x)(x+a)^{-s}$$

$$+ f_1(s) + \int_x^\infty (d/dy)\{(y+a)^{-s}\}B_1(y)dy ,$$

where $f_1(s) = F(s) - f(s)$, by using the partial summation method.

Hence,

$$f_1(s) = \sum_{n\leqq x} a(n)(n+1)^{-s} - \int_1^x B^{(1)}(y)(y+a)^{-s}dy - B(1)(1+a)^{-s}$$

$$- B_1(x)(x+a)^{-s} - \int_x^\infty (d/dy)\{(y+a)^{-s}\}B_1(y)dy ,$$

$$\text{for } \operatorname{Re} s > \beta . \tag{4}$$

By differentiating (4) k times ($k \geqq 0$) with respect to s, we have

$$f_1^{(k)}(s) = \sum_{n\leqq x} a(n)(-\log(n+a))^k(n+a)^{-s}$$

$$- \int_1^x B^{(1)}(y)(-\log(y+a))^k(y+a)^{-s}dy$$

$$- B(1)(-\log(1+a))^k(1+a)^{-s}\delta(a,k)$$

$$- B_1(x)(-\log(x+a))^k(x+a)^{-s}$$

$$- \int_x^\infty (d/dy)\{(-\log(y+a))^k(y+a)^{-s}\}B_1(y)dy ,$$

$$\text{for } \operatorname{Re} s > \beta . \tag{5}$$

The 4-th and 5th terms in the right-hand side of (5) tend to zero as $x \to \infty$. Thus we complete the proof.

< Q.E.D. >

Proof of Theorem 3. From the additional conditions and by the partial integration, (4) becomes to

$$f_1(s) = \sum_{n \leqq x} a(n)(n+a)^{-s} - \int_1^x B^{(1)}(y)(y+a)^{-s}dy - B(1)(1+a)^{-s}$$

$$+ \sum_{k=1}^{N+1} (-1)^k B_k(x)(d^{k-1}/dx^{k-1})(x+a)^{-s}$$

$$+ (-1)^{N+1} \int_x^\infty B_{N+1}(y)(d^{N+1}/dy^{N+1})(y+a)^{-s}dy \ ,$$

$$\text{for } \operatorname{Re} s > \beta - N\delta \ . \qquad (6)$$

By differentiating (6) k times with respect to s we obtain the desired results.

< Q.E.D. >

2. EXAMPLES

Many Dirichlet series appearing in number theory satisfy conditions in Theorem 2 or 3, among which we select two examples.

The First Example: $-\zeta^{(1)}(s)/\zeta(s)$.

$$-\zeta^{(1)}(s)/\zeta(s) = \sum_{n=1}^{\infty} \Lambda(n)n^{-s} \ , \quad \text{for } \operatorname{Re} s > 1 \ ,$$

where $\Lambda(n)$ is von Mangoldt's function.

Then $A(x) = \psi(x) = \sum_{n \leqq x} \Lambda(n) = x + O(x \cdot \exp(-c(\log x)^{1/2}) \ ,$

$$B(x) = x, \ B_1(x) = O(x \cdot \exp(-c(\log x)^{1/2}) \ ,$$

and $f(s) = 1 + (s-1)^{-1}$, which satisfy conditions of Theorem 2 with $\beta = 1$. Hence, we have

$$-\zeta^{(1)}(s)/\zeta(s) = \sum_{k=0}^{\infty} (k!)^{-1}\{f^{(k)}(s_0) + (-1)\delta(k,0)$$

$$+ (-1)^k b(k,s_0)\}(s-s_0)^k, \quad \text{for } \operatorname{Re} s \geqq 1 \ ,$$

and

$$b(k,s_0) = \lim_{x\to\infty} \{ \sum_{n\leqq x} \Lambda(n)(\log n)^k n^{-s_0} - \int_1^x (\log y)^k y^{-s_0} dy\}.$$

The Second Example. $\zeta^2(s)$.

$$\zeta^2(s) = \sum_{n=1}^{\infty} d(n)n^{-s}, \quad \text{for} \quad \text{Re } s > 1 ,$$

where $d(n)$ is the divisor of n and

$$A(x) = D(x) = \sum_{n\leqq x} d(n) = x \log x + (2\gamma-1)x + B_1(x).$$

Then

$$B_1(x) = O(x^{1/3}\log x),$$

$$B_n(x) = O(x^{(n/2)-(1/4)}), \quad \text{for} \quad n \geqq 2,$$

by using the weighted Voronoï formula, and

$$f(s) = (s-1)^{-2} + 2\gamma(s-1)^{-1} + (2\gamma-1),$$

where δ is the Euler constant, which satisfy conditions of Theorem 3 with Re $s_0 > 1/3$.

Then

$$\zeta^2(s) = \sum_{k=0}^{\infty} (k!)^{-1}\{f^{(k)}(s_0) - \delta(k,0)(2\gamma-1) + (-1)^k b(k,s_0)\}(s-s_0)^k,$$

$$b(k,s_0) = \lim_{x\to\infty} \{ \sum_{n\leqq x} d(n)(\log n) n^{-s_0} - \int_1^x (\log y + 2\gamma)(\log y)^k y^{-s_0} dy\},$$

for Re $s_0 > 1/3$.

< Q.E.D. >

REFERENCES

[1] Ivič, A., The Riemann Zeta-function, John Wiley-Sons, New York, 1985.

[2] Mitrovič, D.: Sur la fonction ζ de Riemann, C.R. Acad. Sci., Paris, 244, 1602-1604 (1957); 245, 885-886 (1957).

[3] Nakajima, M.: The Taylor coefficients of $\zeta(s)$, $(s-1)\zeta(s)$ and $(z/(1-z))\zeta(1/(1-z))$, to appear in Math. J. Okayama Univ.

[4] Nakajima, M.: The Taylor coefficients of $\zeta(s)$, preprint.

[5] Briggs, W.E. and Buschman, R.G.: The power series coefficients of functions defined by Dirichlet series, Illinois J. Math., 5, 43-44 (1961).

Proc. Prospects
of Math. Sci.
World Sci. Pub.
189-209, 1988

CRYPTOLOGY - THE MATHEMATICAL THEORY OF DATA SECURITY

HARALD NIEDERREITER*
Mathematical Institute
Austrian Academy of Sciences
Dr. Ignaz-Seipel-Platz 2
A-1010 Vienna
Austria

1. INTRODUCTION

In an age that is more and more dominated by computers and electronic media, the storage and the communication of large amounts of data are vital tasks. Some of these data will be of a confidential nature, e.g. information on private bank accounts, personal medical histories, tax office records, personnel files, university examination questions, and diplomatic and military secrets. These data have to be protected when they are stored in data banks and also when they are transmitted over insecure communication channels. This is done by enciphering the data in such a way that an unauthorized person will find it very difficult to infer the original data. The research area that is concerned with the enciphering and deciphering of information is called *cryptology*. For a long time, cryptology was more an art than a science. Mainly because of skyrocketing demand by commercial users, cryptology has now become an active and vigorous field of inquiry. Particularly the last ten years have seen an enormous increase of research in this area.

Cryptology splits naturally into two parts: the first, *crypto-graphy*, is concerned with designing systems for secure communication

---

* The author gratefully acknowledges support for this work by the Austrian Ministry for Science and Research.

and data protection (i.e. *cryptosystems*) and the second, *cryptanalysis*, deals with breaking cryptosystems. We recall the important distinction between conventional and public-key cryptosystems. A *conventional cryptosystem* consists of an *enciphering scheme* E_k depending on a key k (which is only known to authorized users) and a *deciphering scheme* D_k depending on k. If we are given a *plaintext* m (i.e. data in the original form), then the enciphering scheme produces the *ciphertext* $c = E_k(m)$ and the deciphering scheme recovers m by $D_k(c) = m$. In order to recover m uniquely, we have to require that for fixed k the encryption function E_k is injective. It is a great disadvantage of conventional cryptosystems that we have to know the same secret key k both for enciphering and for deciphering. Therefore this common key k cannot be transmitted over an insecure channel but has to be communicated by some other means, e.g. by a special device called a key transporter. A well-known example of a conventional cryptosystem is the widely used system DES (Data Encryption Standard).

In 1976 Diffie and Hellman [9] came up with the brilliant idea of a *public-key cryptosystem* in which no key exchange is necessary. Each participant in a communication network chooses two keys:

(i) a public key that is made generally available (e.g. in a sort of telephone directory) and is used to encipher plaintext messages sent to the participant;

(ii) a secret (or private) key that is used by the participant to decipher received ciphertexts.

The encryption functions in a public-key cryptosystem must be trapdoor one-way functions. We recall that a function f is *one-way* if f is easy to compute and invertible, but its inverse function f^{-1} is hard to compute, and f is *trapdoor one-way* if f is one-way and f^{-1} becomes easy to compute if additional information (so-called trapdoor information) is known. Ideally, "hard to compute" means that the associated problem belongs to the class of NP-complete problems.

The emphasis in this paper will be on those aspects of cryptology that are currently of greatest interest, namely public-key cryptosystems and stream ciphers. In Section 2 we review some standard examples of cryptosystems, most of them being public-key cryptosystems. In Section 3 we discuss stream ciphers and new results from the theory of linear complexity of sequences. Section 4 is devoted to the FSR cryptosystems introduced by the author [19],[23] and to a general discussion of linear encryption functions. Some public-key cryptosystems of the knapsack type that were designed recently are surveyed in Section 5.

2. SOME STANDARD CRYPTOSYSTEMS

The most widely known public-key cryptosystem is the *RSA cryptosystem* proposed by Rivest, Shamir, and Adleman [27] in 1978. We give only a brief description here since detailed discussions can be found in textbooks on cryptology (see e.g. Konheim [10] and Meyer and Matyas [17]). In the RSA cryptosystem the private key consists of two distinct large primes p and q (each having about 100 decimal digits, say), whereas the public key consists of $n = pq$ and an integer k coprime to $\phi(n) = (p-1)(q-1)$. An acceptable plaintext is an integer m with $0 \leqq m < n$. The corresponding ciphertext is the integer c determined by $c \equiv m^k \pmod{n}$, $0 \leqq c < n$. The legitimate recipient can recover the plaintext m by means of the trapdoor information given by his private key. In the first place he can calculate $\phi(n) = (p-1)(q-1)$, and then he can find a solution d of the congruence $kd \equiv 1 \pmod{\phi(n)}$. Now m is recovered uniquely from

$$c^d \equiv m^{kd} \equiv m \pmod{n}.$$

The security of the RSA system is based on the difficulty of determining $\phi(n)$ from n, or equivalently of factoring n, for very large integers n. Instead of the encryption function $f(x) = x^k \pmod{n}$, one may use certain other polynomials (mod n) or even rational functions (mod n); see e.g. Lidl and Müller [11],[12] and Müller [18].

Another standard example of a public-key cryptosystem is the *knapsack system* of Merkle and Hellman [16]. Here one chooses positive integers $a_1',\cdots,a_n'$ with

$$a'_j > \sum_{i=1}^{j-1} a'_i \quad \text{for} \quad 2 \leq j \leq n,$$

an integer $k > \sum_{i=1}^{n} a'_i$, and an integer w coprime to k. The public key consists of the integers $a_1, \cdots, a_n$ determined by $a_j \equiv wa'_j \pmod{k}$, $0 \leq a_j < k$. All the other data remain private. An acceptable plaintext is a string $m = (m_1, \cdots, m_n) \in \{0,1\}^n$ of n bits. The corresponding ciphertext is the integer $c = \sum_{i=1}^{n} a_i m_i$. Unique decryption is possible on the basis of the private information. A few years after the introduction of the knapsack system, Shamir [30] managed to break this cryptosystem, i.e. he gave an efficient algorithm for determining the private key from the public key. Nevertheless, the basic idea of the knapsack system can be used to construct other cryptosystems of the knapsack type that offer more security; compare with Section 5.

Several cryptosystems are based on the discrete exponential function in finite fields. Let F_q be the finite field of order q and let b be a primitive element of F_q, i.e. a generator of the cyclic multiplicative group F_q^* of F_q. The *discrete exponential function* is the function which assigns to an integer r the element $b^r \in F_q$. Since $b^{q-1} = 1$, it suffices to consider integers r with $0 \leq r \leq q-2$.

The following is a public-key cryptosystem based on the discrete exponential function. The finite field F_q with a very large q and the primitive element b of F_q are known to all the participants in a communication network. A typical participant A chooses an integer h with $2 \leq h \leq q-2$ as the private key and forms the element $b^h \in F_q$ as the public key. The acceptable plaintexts are elements $m \in F_q^*$. If the participant B (with private key k and public key b^k) wants to send the message m to A, then B forms the ciphertext $mb^{hk} = m(b^h)^k$ and transmits it over the channel. The recipient A first computes $b^{hk} = (b^k)^h$ and then recovers the plaintext by calculating $(mb^{hk})(b^{hk})^{-1} = m$. This cryptosystem should only be used for the transmission of a few important messages such as cryptographic keys, since for each pair of participants the plaintext and the ciphertext differ only by a constant factor, which involves a great security risk with repeated use.

A scheme that is designed especially for key exchange is due to Diffie and Hellman [9]. It can be used to set up common keys for conventional cryptosystems such as DES. Suppose again that the finite field F_q with a very large q and the primitive element b of F_q are publicly known. If A and B want to establish a common key, they choose random integers h and k, respectively, with $2 \leqq h, k \leqq q-2$. Then A sends b^h to B, while B transmits b^k to A. They both take b^{hk} as their common key, which can be computed by A as $(b^k)^h$ and by B as $(b^h)^k$.

Discussions of cryptosystems based on finite fields can be found in Lidl and Niederreiter [14, Ch.9] and Niederreiter [22]. We mention also a recent proposal by Japanese cryptologists (see Tsujii et al. [32]) of a public-key cryptosystem based on finite field arithmetic.

3. STREAM CIPHERS

A simple conventional cryptosystem which offers a high level of security is the *one-time pad*. Here the acceptable plaintexts are strings of bits, where each bit is regarded as an element of F_2, the finite field of order 2. The key is a random string of bits and the ciphertext is obtained by bit-by-bit addition of plaintext and key, using addition in F_2. The plaintext is recovered by bit-by-bit addition of ciphertext and key, using again addition in F_2 and the fact that $a + a = 0$ for all $a \in F_2$. If the key string is truly random, then the ciphertext string is also truly random and so it provides no clue for persons who do not possess the secret key.

Some of the disadvantages of the one-time pad are: (i) it requires very long keys, since the key string must be at least as long as the message string and is only used once; (ii) the difficulty of generating truly random key strings and transporting them from sender to receiver. In practice one works with a *stream cipher*, in which the random key strings of bits are replaced by pseudorandom sequences of bits that can be generated deterministically from shorter keys. The crucial issue here is to decide when a pseudorandom sequence has satisfactory randomness properties from the cryptographic viewpoint, which means that it should be difficult to infer the parameters (= keys) in the algorithm generating the sequence.

The building blocks in many methods of generating pseudorandom sequences are linear recurring (or feedback shift register) sequences in finite fields. In the applications to stream ciphers we are only interested in F_2, but the theory can be developed for any finite field F_q. Let $k \geq 1$ be an integer. A sequence (s_i), $i = 1,2,\cdots$, of elements of F_q is called a *kth-order linear recurring* (or *feedback shift register) sequence* if it satisfies a kth-order linear recurrence relation

$$s_{i+k} = a_{k-1}s_{i+k-1}+\cdots+a_1s_{i+1}+a_0s_i \quad \text{for} \quad i = 1,2,\cdots \tag{1}$$

with constant coefficients $a_0,a_1,\cdots,a_{k-1} \in F_q$. The sequence (s_i) is completely determined by (1) and by the initial values $s_1,s_2,\cdots,$ s_k. A linear recurring sequence can easily be generated by a simple switching circuit called a feedback shift register, hence it is the synonym of feedback shift register sequence. A kth-order linear recurring sequence itself is not suitable as a pseudorandom sequence since any $2k$ consecutive terms of the sequence determine the coefficients in (1) and the initial values, and hence the whole sequence (see [13, Ch.8], [14, Ch.6]). Therefore one uses various devices, such as multiplexing, to combine several linear recurring sequences in order to obtain more complex sequences; see [14, Ch.9] and the relevant articles in [25].

In this context, a useful measure for complexity is the linear complexity. If (s_i) is an arbitrary sequence of elements of F_q and n is a positive integer, then the *linear complexity* $L(n)$ of (s_i) is defined as the least k such that the initial segment

$s_1, s_2, \cdots, s_n$ can be generated by a kth-order linear recurrence relation (1), with the provision that $L(n) = 0$ if $s_i = 0$ for $1 \leqq i \leqq n$. The linear complexity can be efficiently calculated by the Berlekamp-Massey algorithm (see [13, Ch.8], [15]). It is clear that for fixed (s_i) the linear complexity $L(n)$ is a nondecreasing function of n. The sequence $L(1), L(2), \cdots$ is called the *linear complexity profile* of (s_i).

Taking account of a definition for $q = 2$ given by Rueppel [28], [29, Ch.4] and Wang and Massey [33], we say that the sequence (s_i) of elements of F_q has a *perfect linear complexity profile* (PLCP) if

$$L(n) = \left\lfloor \frac{n+1}{2} \right\rfloor \quad \text{for all } n \geqq 1,$$

where $\lfloor t \rfloor$ denotes the greatest integer $\leqq t$. This definition stems from the result of Rueppel [28],[29, Ch.4] that, for random sequences of elements of F_2, the expected value of $L(n)$ is $\frac{n}{2} + c_n$ with $0 \leqq c_n \leqq \frac{5}{18}$.

There is a close connection between the linear complexity profile of a sequence and the continued fraction expansion of an associated generating function. In particular, the sequences with a PLCP can be characterized completely in terms of the continued fraction expansion of the generating function. If (s_i) is an arbitrary sequence of elements of F_q, then we associate with it the *generating function* $\sum_{i=1}^{\infty} s_i x^{-i}$, regarded as an element of the ring $F_q[[x^{-1}]]$ of formal power series over F_q in x^{-1}. The linear recurring sequences are exactly those whose generating function is rational (see Lemma 1 below). We recall that if (s_i) is a linear recurring sequence defined by (1), then

$$f(x) = x^k - a_{k-1}x^{k-1} - \cdots - a_1 x - a_0 \in F_q[x] \qquad (2)$$

is called the *characteristic polynomial* of (s_i). By definition, the constant polynomial 1 is also regarded as the characteristic polynomial of the zero sequence. As usual we define $\deg(0) = -\infty$.

Lemma 1. *Let* $f \in F_q[x]$ *be monic and let* (s_i) *be a sequence of elements of* F_q. *Then* (s_i) *is a linear recurring sequence with characteristic polynomial* f *if and only if*

$$\sum_{i=1}^{\infty} s_i x^{-i} = \frac{h(x)}{f(x)}$$

with $h \in F_q[x]$ *and* $\deg(h) < \deg(f)$.

Proof. This is clear if $f = 1$. If f is given as in (2) with degree $k \geq 1$, then consider

$$f(x) \sum_{i=1}^{\infty} s_i x^{-i} = (x^k - a_{k-1}x^{k-1} - \cdots - a_1 x - a_0)(s_1 x^{-1} + s_2 x^{-2} + \cdots).$$

The right-hand side is a polynomial of degree $<k$, if and only if the coefficient of each x^j, $j < 0$, vanishes, which means that

$$s_{-j+k} - a_{k-1}s_{-j+k-1} - \cdots - a_1 s_{-j+1} - a_0 s_{-j} = 0 \quad \text{for all} \quad j < 0,$$

and this is equivalent to the validity of (1).

< Q.E.D. >

If (s_i) is a linear recurring sequence, then its rational generating function has a uniquely determined reduced form h/m with $h, m \in F_q[x]$, $\deg(h) < \deg(m)$, $\gcd(h,m) = 1$, and m monic. The polynomial m is called the *minimal polynomial* of (s_i). It can also be described as the characteristic polynomial of (s_i) of least degree.

Lemma 2. *Let* $m \in F_q[x]$ *be monic and let* (s_i) *be a sequence of elements of* F_q. *Then* (s_i) *is a linear recurring sequence with minimal polynomial* m *if and only if*

$$\sum_{i=1}^{\infty} s_i x^{-i} = \frac{h(x)}{m(x)}$$

with $h \in F_q[x]$, $\deg(h) < \deg(m)$, *and* $\gcd(h,m) = 1$.

Proof. This follows from Lemma 1 and the definition of the minimal polynomial.

< Q.E.D. >

It is clear that Lemmas 1 and 2 also hold for arbitrary fields. However, the following characterization is typical for finite fields: (s_i) is a linear recurring sequence if and only if it is ultimately periodic (compare with [13, p.437]).

On the field $F_q(x)$ of rational functions over F_q a non-archimedean valuation $|\ |$ is defined by

$$\left|\frac{f}{g}\right| = 2^{\deg(f)-\deg(g)} \quad \text{for } f,g \in F_q[x],\ g \neq 0.$$

This valuation is extended to the valuation on the quotient field of $F_q[[x^{-1}]]$ by setting:

$$|S| = 2^{-r} \quad \text{if } S = \sum_{i=r}^{\infty} s_i x^{-i} \quad \text{and } s_r \neq 0, \text{ and } |0| = 0.$$

Every irrational $S = \sum_{i=1}^{\infty} s_i x^{-i} \in F_q[[x^{-1}]]$ has a unique *continued fraction expansion*

$$S = 1/(A_1 + 1/(A_2 + \cdots)) = :[A_1, A_2, \cdots]$$

with partial quotients $A_j \in F_q[x]$ and $\deg(A_j) \geq 1$ for $j \geq 1$. If we break off the expansion after the term A_j, $j \geq 1$, we get the rational convergent P_j/Q_j. The polynomials P_j and Q_j can be calculated recursively by

$$\begin{aligned} P_{-1} &= 1,\ P_0 = 0,\ P_j = A_j P_{j-1} + P_{j-2} \quad \text{for } j \geq 1, \\ Q_{-1} &= 0,\ Q_0 = 1,\ Q_j = A_j Q_{j-1} + Q_{j-2} \quad \text{for } j \geq 1. \end{aligned} \tag{3}$$

It follows that

$$\deg(Q_j) = \sum_{h=1}^{j} \deg(A_h) \quad \text{for } j \geq 1. \tag{4}$$

We also have $\gcd(P_j, Q_j) = 1$ for $j \geq 0$ and

$$|Q_j S - P_j| = |Q_{j+1}|^{-1} = |A_{j+1}|^{-1}\, |Q_j|^{-1} \quad \text{for } j \geq 0. \tag{5}$$

These standard facts can be found e.g. in de Mathan [8], where the following result is also shown.

Lemma 3. *If* $|fS - g| \leq |f|^{-1}$ *for some* $f,g \in F_q[x]$ *with* $\gcd(f,g) = 1$ *and* $f \neq 0$, *then for some* $j \geq 0$ *and* $c,d \in F_q$ *with* $c \neq 0$ *we have*

$$f = cQ_j + dQ_{j-1}, \quad g = cP_j + dP_{j-1}.$$

If $|fS - g| < |f|^{-1}$, *then* $d = 0$.

We can now establish the connection between linear complexity profiles and continued fractions. For irrational $S = \sum_{i=1}^{\infty} s_i x^{-i} \in F_q[[x^{-1}]]$ with partial quotients $A_1, A_2, \cdots$, we define

$$K(S) = \sup_{j \geq 1} \deg(A_j),$$

where we can have $K(S) = \infty$. We note that S is irrational if and only if (s_i) is not a linear recurring sequence.

Theorem 1. *If the sequence* (s_i) *of elements of* F_q *is such that* $S = \sum_{i=1}^{\infty} s_i x^{-i} \in F_q[[x^{-1}]]$ *is irrational, then the linear complexity* $L(n)$ *of* (s_i) *satisfies*

$$\frac{1}{2}(n + 1 - K(S)) \leq L(n) \leq \frac{1}{2}(n + K(S)) \quad \textit{for all } n \geq 1.$$

Proof. We first prove the lower bound. If, for fixed $n \geq 1$, we have $L(n) \geq (n + 1)/2$, then the lower bound is trivial, so we can assume $L(n) < (n + 1)/2$. The definition of $L(n)$ shows that there exists a linear recurring sequence (t_i) with minimal polynomial $m_n \in F_q[x]$ of degree $L(n)$ and $t_i = s_i$ for $1 \leq i \leq n$. Lemma 2 implies that for a suitable $h_n \in F_q[x]$ with $\gcd(h_n, m_n) = 1$ we have

$$\frac{h_n}{m_n} = \sum_{i=1}^{\infty} t_i x^{-i} = S + x^{-n-1} R_n$$

with $0 < |R_n| \leq 1$, hence

$$|m_n S - h_n| = |- x^{-n-1} m_n R_n| = |m_n|^{-1}\, 2^{2L(n)-n-1}\, |R_n|.$$

Since $2L(n)-n-1 < 0$, we can apply Lemma 3, and together with (5) this yields

$$|A_{j(n)+1}| = 2^{n+1-2L(n)} \, |R_n|^{-1} \geq 2^{n+1-2L(n)}$$

for some partial quotient $A_{j(n)+1}$ of S with $j(n) \geq 0$. Therefore

$$K(S) \geq \deg(A_{j(n)+1}) \geq n + 1 - 2L(n),$$

which implies the desired lower bound for $L(n)$.

To prove the upper bound, we note that (5) yields for any $j \geq 0$,

$$Q_j S - P_j = x^{-\deg(Q_{j+1})} U_j$$

with $|U_j| = 1$, or equivalently

$$Q_j(S - x^{-\deg(Q_j)-\deg(Q_{j+1})} T_j) = P_j$$

with $|T_j| = 1$. Furthermore

$$S - x^{-\deg(Q_j)-\deg(Q_{j+1})} T_j = \sum_{i=1}^{\infty} t_i x^{-i}$$

with $t_i = s_i$ for $1 \leq i \leq \deg(Q_j) + \deg(Q_{j+1}) - 1$, hence

$$\sum_{i=1}^{\infty} t_i x^{-i} = \frac{P_j}{Q_j} .$$

Now $\deg(P_j) < \deg(Q_j)$ by (3) and $\gcd(P_j, Q_j) = 1$, so it follows from Lemma 2 that (t_i) is a linear recurring sequence with minimal polynomial of degree $\deg(Q_j)$. Thus

$$L(n) \leq \deg(Q_j) \quad \text{for } 1 \leq n \leq \deg(Q_j) + \deg(Q_{j+1}) - 1. \tag{6}$$

Now let $n \geq 1$ be given and choose the least $j \geq 0$ such that $n \leq \deg(Q_j) + \deg(Q_{j+1}) - 1$. If $j = 0$, then (6) implies $L(n) = 0$ and the upper bound is valid. If $j \geq 1$, then

$$n \geq \deg(Q_{j-1}) + \deg(Q_j) = 2\deg(Q_j) - \deg(A_j) \geq 2\deg(Q_j) - K(S),$$

and so by (6) we get $L(n) \leq \deg(Q_j) \leq \frac{1}{2}(n + K(S))$.

< Q.E.D. >

Theorem 2. *If the sequence* (s_i) *of elements of* F_q *is such that* $S = \sum_{i=1}^{\infty} s_i x^{-i} \in F_q[[x^{-1}]]$ *is irrational, then*

$$K(S) = \sup_{n \geq 1} (n + 1 - 2L(n)). \tag{7}$$

Proof. Let M be the right-hand side of (7). For $j \geq 0$ we put $n_j = \deg(Q_j) + \deg(Q_{j+1}) - 1$. If $n_j \geq 1$, then $L(n_j) \leq \deg(Q_j)$ by (6), and so

$$M \geq n_j + 1 - 2L(n_j) \geq \deg(Q_j) + \deg(Q_{j+1}) - 2\deg(Q_j) = \deg(A_{j+1}). \tag{8}$$

This shows in particular that $M \geq 1$. If $n_j = 0$, then necessarily $j = 0$ and $\deg(Q_1) = \deg(A_1) = 1$, and so (8) holds again. It follows from (8) that $M \geq K(S)$. The reverse inequality is obtained from the lower bound in Theorem 1.

< Q.E.D. >

Corollary 1. *If the linear complexity* $L(n)$ *of the sequence* (s_i) *of elements of* F_q *satisfies* $L(n) \geq n/2$ *for all* $n \geq 1$, *then* $S = \sum_{i=1}^{\infty} s_i x^{-i} \in F_q[[x^{-1}]]$ *is irrational and* $K(S) = 1$.

Proof. From $L(n) \geq n/2$ for all $n \geq 1$ we get $\lim_{n\to\infty} L(n) = \infty$, hence S is irrational. The fact that $K(S) = 1$ follows from Theorem 2.

< Q.E.D. >

One can show that if $L(n) \leq (n+1)/2$ for all $n \geq 1$ and $\lim_{n\to\infty} L(n) = \infty$, then the conclusions of Corollary 1 still hold. The proof proceeds as in [21], where the case $q = 2$ was treated.

Theorem 3. *The sequence* (s_i) *of elements of* F_q *has a perfect linear complexity profile if and only if* $S = \sum_{i=1}^{\infty} s_i x^{-i} \in F_q[[x^{-1}]]$ *is irrational and* $K(S) = 1$.

Proof. If (s_i) has a PLCP, then Corollary 1 shows that S is irrational and $K(S) = 1$. Conversely, if S is irrational and $K(S) = 1$, then Theorem 1 implies that $n/2 \leq L(n) \leq (n+1)/2$, hence

$L(n) = \lfloor (n+1)/2 \rfloor$ for all $n \geqq 1$, and so (s_i) has a PLCP.

< Q.E.D. >

In the case $q = 2$, the irrational $S \in F_2[[x^{-1}]]$ with $K(S) = 1$ has been characterized by Baum and Sweet [1], and together with Theorem 3, this leads to the result that a binary sequence (s_i) has a PLCP if and only if $s_1 = 1$ and $s_{2i+1} = s_{2i} + s_i$ for all $i \geqq 1$. This characterization was first shown by Wang and Massey [33] via a different argument. A special class of binary sequences with a PLCP was studied in Niederreiter [21].

The characterization of binary sequences with a PLCP indicates that, for such sequences, all the terms s_i with odd index i can be calculated from previous terms, and so these sequences are not suitable as pseudorandom sequences. Therefore one should consider sequences with an "almost perfect" linear complexity profile, i.e. sequences in which slight deviations of $L(n)$ from its expected value are allowed. Such sequences can be constructed on the basis of Theorem 2, by considering generating functions S with a relatively small $K(S) > 1$. For instance, choose $S \in F_2[[x^{-1}]]$ with $K(S) = 2$ or 3, that would be in accordance with the formula of Rueppel [28], [29,Ch.4] for the variance of $L(n)$ for random binary sequences.

For a different approach to the construction of pseudorandom sequences for cryptographic purposes, we refer to Blum et al. [3] and Blum and Micali [4].

4. FSR CRYPTOSYSTEMS

In Section 2 we discussed some cryptosystems based on the discrete exponential function in finite fields. These cryptosystems can be broken if the following *discrete logarithm problem* can be solved efficiently: given a primitive element $b \in F_q$ and a $c \in F_q^*$, find the uniquely determined integer r, $0 \leqq r \leqq q-2$, for which $c = b^r$. In recent years several algorithms have been developed which solve the discrete logarithm problem in subexponential time. We refer to the surveys in Lidl and Niederreiter [14, Ch.9] and Odlyzko [24] and to the papers of Coppersmith [6] and Coppersmith et al. [7]. These discrete logarithm algorithms have reduced the security of cryptosystems based on the discrete exponential function.

The author [19],[23] has proposed cryptosystems that employ a more complex operation than discrete exponentiation, namely the decimation of linear recurring (or feedback shift register, abbreviated FSR) sequences. If (s_i), $i = 0,1,\cdots$, is a sequence of elements of F_q, then the *decimation* of (s_i) by the factor $k \geqq 1$ yields the sequence (s_{ik}), $i = 0,1,\cdots$. If (s_i) is an nth-order FSR sequence, then any decimation produces again an nth-order FSR sequence (see [23, Theorem 1]).

The following is a conventional cryptosystem based on the decimation of FSR sequences. The acceptable plaintexts are strings of length $n \geqq 2$ of elements of the finite field F_q of characteristic p. The key is a random integer k with

$$1 < k < R: = p^u \operatorname{lcm}(q - 1, q^2 - 1,\cdots,q^n - 1) \tag{9}$$

and $\gcd(k,R) = 1$, where u is the least integer with $p^u \geqq n$. If A wants to send the plaintext message $a_0a_1\cdots a_{n-1}$ to B, then A forms the polynomial

$$f(x) = x^n - a_{n-1}x^{n-1}-\cdots-a_1x - a_0 \in F_q[x].$$

Let (s_i), $i = 0,1,\cdots$, be the FSR sequence with characteristic polynomial f and initial values $s_0 = s_1 =\cdots=s_{n-2} = 0$, $s_{n-1} = 1$. Then the ciphertext is the string $s_ks_{2k}\cdots s_{(2n-1)k}$ of $2n - 1$ elements of F_q. For decryption B considers the decimated sequence $(t_i) = (s_{ik})$. Then B knows t_i for $1 \leqq i \leqq 2n - 1$ and also $t_0 = s_0 = 0$, and so the first $2n$ terms of (t_i). The minimal polynomial of (t_i) has degree $\leqq n$, hence it can be calculated by the Berlekamp-Massey algorithm and the rest of the sequence (t_i) is determined. For $0 \leqq j \leqq n - 1$ let $d_j \geqq 1$ be a solution of the congruence $kd_j \equiv n + j \pmod R$. Then

$$s_{n+j} = s_{kd_j} = t_{d_j} \quad \text{for } 0 \leqq j \leqq n - 1,$$

and so B knows the first $2n$ terms of the sequence (s_i). On the basis of this information B can calculate f and thus recover the plaintext.

The following is an analog of the Diffie-Hellman key-exchange system. Suppose the nth-order FSR sequence (s_i), $i = 0,1,\cdots,$ and the positive integer $m \leqq 2n - 1$ are publicly known. If A and B want to establish a common key, they choose random integers h and k, respectively, with $1 < h, k < e$, where e is the least period of (s_i). Then A sends the string $s_0 s_h s_{2h} \cdots s_{(2n-1)h}$ to B, while B sends the string $s_0 s_k s_{2k} \cdots s_{(2n-1)k}$ to A. Now A and B can calculate the minimal polynomials of $(t_i) = (s_{ik})$ and $(u_i) = (s_{ih})$, respectively. Since $t_{ih} = s_{ihk} = u_{ik}$ for all $i \geqq 0$, both A and B have the same key $t_h t_{2h} \cdots t_{mh}$ consisting of a string of m elements of F_q.

A variant of Shamir's no-key algorithm (see [10, pp.345-346]) can be set up as follows. Suppose A wants to send a message to B that consists of a string $a_0 a_1 \cdots a_{n-1}$ of $n \geqq 2$ elements of F_q. Then A chooses a random integer h with $1 < h < R$ and $\gcd(h,R) = 1$, where R is the same as in (9). Now A considers the FSR sequence (s_i), $i = 0,1,\cdots,$ with initial values $s_0 = s_1 = \cdots = s_{n-2} = 0$, $s_{n-1} = 1$, and characteristic polynomial

$$f(x) = x^n - a_{n-1}x^{n-1} - \cdots - a_1 x - a_0 \in F_q[x],$$

and sends the string $s_h s_{2h} \cdots s_{(2n-1)h}$ to B. On the basis of this information, B calculates the minimal polynomial of the decimated sequence $(t_i) = (s_{ih})$. Then B chooses a random integer k with $1 < k < R$ and $\gcd(k,R) = 1$ and transmits the string $t_k t_{2k} \cdots t_{(2n-1)k}$ to A. On the basis of this information, A calculates the minimal polynomial of the decimated sequence $(u_i) = (t_{ik})$. Furthermore, A determines a solution $m \geqq 1$ of the congruence $hm \equiv 1 \pmod{R}$ and sends the string $u_m u_{2m} \cdots u_{(2n-1)m}$ to B. This allows B to calculate the minimal polynomial of the decimated sequence $(v_i) = (u_{im})$. Then B determines for $0 \leqq j \leqq n - 1$ a solution $d_j \geqq 1$ of the congruence $kd_j \equiv n + j \pmod{R}$. Then

$$s_{n+j} = v_{d_j} \quad \text{for } 0 \leq j \leq n-1,$$

so that B knows the first $2n$ terms of the sequence (s_i) and thus can recover the plaintext.

The following is a public-key cryptosystem based on FSR sequences. Each participant in the communication network knows the FSR sequence (s_i) with minimal polynomial $g \in F_q[x]$, where g has degree $n \geq 2$ and is not divisible by x^2. A typical participant A chooses as the private key an integer h with $1 < h < e$ and $\gcd(h,e) = 1$, where e is the least period of (s_i), and the corresponding public key consists of the minimal polynomial g_h and the initial values $t_0, t_1, \cdots, t_{n-1}$ of the decimated sequence $(t_i) = (s_{ih})$. Acceptable plaintext messages are nonzero row vectors $\underline{m} \in F_q^n$. If a second participant B with private key k wants to send the message $\underline{m}$ to A, then B sets up the Hankel matrix

$$V = \begin{bmatrix} v_0 & v_1 \cdots v_{n-1} \\ v_1 & v_2 \cdots v_n \\ \cdot & \cdot & \cdot \\ \cdot & \cdot & \cdot \\ \cdot & \cdot & \cdot \\ v_{n-1} & v_n \cdots v_{2n-2} \end{bmatrix}$$

with $v_i = t_{ik}$ and calculates the vector $\underline{c} = \underline{m}V$ as the ciphertext. To decipher $\underline{c}$, A uses the knowledge of the elements $u_i = s_{ik}$ from B's public key to calculate $v_i = s_{ihk} = u_{ih}$. Thus A finds the matrix V, which can be shown to be nonsingular, and so A recovers $\underline{m} = \underline{c}V^{-1}$.

The security of the FSR public-key cryptosystem is based on the difficulty of calculating h, given the polynomials g and g_h. If g is fractorized in its splitting field over F_q into

$$g(x) = \prod_{j=1}^{n} (x - \beta_j),$$

then, under the conditions above, we have

$$g_h(x) = \prod_{j=1}^{n} (x - \beta_j^h).$$

In the case where g is irreducible over F_q, the problem of determining h is equivalent to a discrete logarithm problem; this comment was already made in Niederreiter [19] and is repeated in Smeets [31]. Therefore it is advisable to choose g in such a way that it factors into many polynomials over F_q of small degree, with q being large.

The encryption function in the FSR public-key cryptosystem is linear. In general, for a linear encryption function an acceptable plaintext is a row vector $\underline{m} \neq \underline{0}$ and the corresponding ciphertext is $\underline{m}E$, where E is a suitable matrix. The matrix E may be known (i.e., equal to the public key), e.g. in knapsack systems, or it may be unknown (i.e., constructed from secret data), e.g. in the FSR public-key cryptosystem. In both cases the cryptosystem is unsafe for plaintexts of small (Hamming) weight. For instance, if $\underline{m} = (0,\cdots,0,1,0,\cdots,0)$, where the component 1 is in the jth position, then $\underline{m}E$ is equal to the jth row of E. If E is known, then the opponent immediately knows $\underline{m}$; if E is unknown, then the opponent can use this $\underline{m}$ in a chosen-plaintext attach to get information on E. A simple technique of avoiding low-weight plaintexts is the following. Choose a linear (n,k) code with length n equal to the length of $\underline{m}$, dimension $k < n$, and a large minimum distance d. Let G be a generator matrix of the code, i.e. a $k \times n$ matrix whose row space is equal to the code. The plaintexts are now row vectors $\underline{x} \neq \underline{0}$ of length k and the corresponding ciphertexts are given by $\underline{x}GE$. We note that each code word $\underline{x}G$ has weight $\geq d$, thus no low-weight words enter the cryptosystem. For decryption we first invert the encryption function E and then the coding scheme.

FSR cryptosystems offer the possibility of *error-detecting cryptography*. In all these cryptosystems, strings of consecutive terms of certain FSR sequences are transmitted over the communication channel. A few subsequent terms of the FSR sequence can be added as check symbols in order to detect transmission errors. If the check symbols do not fit the recurrence relation, the receiver can ask for a retransmission.

5. KNAPSACK-TYPE CRYPTOSYSTEMS

A *knapsack-type cryptosystem* is a public-key cryptosystem based on the difficulty of determining the coefficients in a linear combination from the value of the linear combination. As we pointed out in Section 2, the classical knapsack system has been broken, but there are knapsack-type cryptosystems that offer more security. An obvious variant of the classical system is obtained by changing the ring of integers into a different ring; with polynomial rings over finite fields this has been carried out by Pieprzyk [26].

A knapsack-type cryptosystem designed for low-weight plaintexts was introduced by Chor and Rivest [5]. A description of this cryptosystem can also be found in Lidl and Niederreiter [14, Ch.9]. It is a serious disadvantage of this cryptosystem that it has a low information rate, i.e. the information per bit of ciphertext is small. A knapsack-type cryptosystem for low-weight plaintexts which yields a significantly higher information rate was proposed by the author [20]. Like the well-known Goppa-code cryptosystem (see [14, Ch.9]), it is based on ideas from algebraic coding theory.

We now describe this knapsack-type cryptosystem in more detail. The finite field F_q and the parameters n, k, and t are known to each participant in the communication network. A typical participant A chooses a t-error-correcting linear (n,k) code C over F_q and a parity-check matrix H of C, i.e. an $(n-k) \times n$ matrix H of rank $n-k$ such that C is the null space of H. Furthermore, A chooses a non-singular $(n-k) \times (n-k)$ matrix M over F_q and an $n \times n$ matrix P over F_q obtained by permuting the rows of a nonsingular diagonal matrix. The matrices M, H, and P form the private key, whereas the $(n-k) \times n$ matrix $K = MHP$ serves as the public key. Acceptable plaintexts are column vectors $\underline{m} \in F_q^n$ of weight $\leq t$. The ciphertext corresponding to $\underline{m}$ is the vector $\underline{c} = K\underline{m}$ which is sent to A. Since $\underline{c} = MHP\underline{m}$, the recipient A can premultiply $\underline{c}$ by M^{-1} to get $H(P\underline{m})$. We note that $P\underline{m}$ is a vector of weight $\leq t$, hence an application of the decoding algorithm of C to the syndrome $H(P\underline{m})$ yields $P\underline{m}$. Then A recovers $\underline{m}$ by premultiplying $P\underline{m}$ by P^{-1}. The security of this cryptosystem is based on the following facts: (i) to an opponent the matrix K looks like the parity-check matrix of a random linear (n,k) code; (ii) the problem of decoding for a random linear code is NP-complete by a result of Berlekamp et al. [2].

It is clear that the codes C should be chosen in such a way that they have a relatively large error-correcting capability and that they allow efficient decoding algorithms. To get a high information rate, the codes C should asymptotically meet the Gilbert-Varshamov bound. We refer to [20] for a more detailed discussion of the choice of codes. As pointed out by A.M. Odlyzko, in the concrete examples given in [20] one should work with codes having a larger dimension k. The reason is that for a given ciphertext $\underline{c}$, the equation $K\underline{m} = \underline{c}$ has q^k solutions $\underline{m}$, and if k is too small, one can search all these solutions until one finds the unique solution of weight $\leq t$.

REFERENCES

[1] Baum, L.E. and Sweet, M.M.: Badly approximable power series in characteristic 2, Ann. of Math., 105, 573-580 (1977).

[2] Berlekamp, E.R., McEliece, R.J. and van Tilborg, H.: On the inherent intractability of certain coding problems, IEEE Trans. Information Theory, 24, 384-386 (1978).

[3] Blum, L., Blum, M. and Shub, M.: A simple unpredictable pseudo-random number generator, SIAM J. Computing, 15, 364-383 (1986).

[4] Blum, M. and Micali, S.: How to generate cryptographically strong sequences of pseudo-random bits, SIAM J. Computing, 13, 850-864 (1984).

[5] Chor, B. and Rivest, R.L.: A knapsack type public key crypto-system based on arithmetic in finite fields, Advances in Cryptology (Proc. CRYPTO 84, Santa Barbara, 1984), 54-65, Lecture Notes in Computer Science, 196, Springer-Verlag, Berlin, 1985.

[6] Coppersmith, D.: Fast evaluation of logarithms in fields of characteristic two, IEEE Trans. Information Theory, 30, 587-594 (1984).

[7] Coppersmith, D., Odlyzko, A.M. and Schroeppel, R.: Discrete logarithms in GF(p), Algorithmica, 1, 1-15 (1986).

[8] de Mathan, B.: Approximations diophantiennes dans un corps local, Bull. Soc. Math. France Suppl. Mém., 21 (1970).

[9] Diffie, W. and Hellman, M.E., New directions in cryptography, IEEE Trans. Information Theory, 22, 644-654 (1976).

[10] Konheim, A.G., Cryptography. A Primer, Wiley, New York, 1981.

[11] Lidl, R. and Müller, W.B.: Permutation polynomials in RSA-cryptosystems, Proc. CRYPTO '83 (Santa Barbara, 1983), 293-301, Plenum, New York, 1984.

[12] Lidl, R. and Müller, W.B.: A note on polynomials and functions in algebraic cryptography, Ars Combinatoria, 17A, 223-229 (1984).

[13] Lidl, R. and Niederreiter, H., Finite Fields, Addison-Wesley, Reading, 1983.

[14] Lidl, R. and Niederreiter, H., Introduction to Finite Fields and Their Applications, Cambridge Univ. Press, Cambridge, 1986.

[15] Massey, J.L.: Shift-register synthesis and BCH decoding, IEEE Trans. Information Theory, 15, 122-127 (1969).

[16] Merkle, R.C. and Hellman, M.E.: Hiding information and signatures in trapdoor knapsacks, IEEE Trans. Information Theory, 24, 525-530 (1978).

[17] Meyer, C.H. and Matyas, S.M., Cryptography. A New Dimension in Computer Data Security, Wiley, New York, 1982.

[18] Müller, W.B.: Polynomial functions in modern cryptology, Contributions to General Algebra, 3 (Proc. Conf. Vienna, 1984), 7-32, Teubner, Stuttgart, 1985.

[19] Niederreiter, H.: A public-key cryptosystem based on shift register sequences, in [25], 35-39.

[20] Niederreiter, H.: Knapsack-type cryptosystems and algebraic coding theory, Problems of Control and Information Theory, 15, 159-166 (1986).

[21] Niederreiter, H., Continued fractions for formal power series, pseudo-random numbers and linear complexity of sequences, Contributions to General Algebra, 5 (Proc. Conf. Salzburg, 1986), 221-233, Teubner, Stuttgart, 1987.

[22] Niederreiter, H.: Algebraische Methoden zum Entwurf kryptographischer Systeme, Elektrotechnik und Maschinenbau, to appear.

[23] Niederreiter, H.: Some new cryptosystems based on feedback shift register sequences, Math. J. Okayama Univ., to appear.

[24] Odlyzko, A.M.: Discrete logarithms in finite fields and their cryptographic significance, Advances in Cryptology (Proc. EUROCRYPT 84, Paris, 1984), 224-314, Lecture Notes in Computer Science, 209, Springer-Verlag, Berlin, 1985.

[25] Pichler, F. (ed.), Advances in Cryptology-EUROCRYPT '85 (Linz, 1985), Lecture Notes in Computer Science, 219, Springer-Verlag, Berlin, 1986.

[26] Pieprzyk, J.P.: On public-key cryptosystems built using polynomial rings, in [25], 73-78.

[27] Rivest, R.L., Shamir, A. and Adleman, L.: A method for obtaining digital signatures and public-key cryptosystems, Comm. ACM, 21, 120-126 (1978).

[28] Rueppel, R.A.: Linear complexity and random sequences, in [25], 167-188.

[29] Rueppel, R.A., Analysis and Design of Stream Ciphers, Springer-Verlag, Berlin, 1986.

[30] Shamir, A.: A polynomial time algorithm for breaking the basic Merkle-Hellman cryptosystem, Proc. 23rd Symp. on Foundations of Computer Science (Chicago, 1982), 145-152, IEEE Computer Society, Los Angeles, 1982.

[31] Smeets, B.: A comment on Niederreiter's public key cryptosystem, in [25], 40-42.

[32] Tsujii, S., Kurosawa, K., Itoh, T., Fujioka, A. and Matsumoto, T.: A public-key cryptosystem based on the difficulty of solving a system of nonlinear equations, Tech. Memorandum, 1, Tsujii Laboratory, Tokyo Institute of Technology, 1986.

[33] Wang, M.Z. and Massey, J.L.: The characterization of all binary sequences with a perfect linear complexity profile, paper presented at EUROCRYPT '86 (Linköping, 1986).

Proc. Prospects
of Math. Sci.
World Sci. Pub.
211-230, 1988

DIVISOR PROBLEMS AND EXPONENT PAIRS: ON A CONJECTURE BY CHOWLA AND WALUM.

Y.-F.S. PETERMANN*

Université de Genève
Section de Mathématiques
2-4, rue du Lièvre, C.P. 240
CH-1211 Genève 24
Switzerland

0. INTRODUCTION AND ACKNOWLEDGEMENTS

The essential part of this article is devoted to the discussion of two items. One is a conjecture of Chowla and Walum generalizing the famous Piliz-Hardy-Landau conjecture on Dirichlet's divisor problem (Section 2), which we introduce in Section 3 through a brief historical account of one of the classical divisor problems. The other is the very elegant theory of one-dimensional exponent pairs invented by Van der Corput, simplified by Phillips, and investigated further by Rankin and S.W. Graham (Section 4). The results stated in Section 5 connect the material in Sections 3 and 4.

In Section 1 an explanation is proposed for a fact that I used to find somewhat puzzling: there is no generally accepted-nor acceptable-standard definition of the error term arising from a summatory function.

The statements of original results are included in this work; the proofs will appear elsewhere: (1.10) in [28] (also see [26]), (1.18) in [30], and the theorems of Section 5 in [29]. Also, improvements on some recent results of Graham's are given in Section 4A.

---

* 1980 AMS classification numbers: primary 10H25; secondary 10G10.

Acknowledgements

This work has been done while I was 1) supported by the JSPS-SNSF exchange program and the Fonds Marc Birkigt (Genève), and visiting Kyushu University in Fukuoka; 2) supported by an SNSF fellowship and visiting the University of Illinois at Urbana-Champaign. The efficient typing was provided by the Mathematics Departments of the University of Illinois and the University of Genève. I am grateful to these institutions for their help.

1. ERROR TERMS FOR SUMMATORY FUNCTIONS

There is a popular concern in analytic number theory, which consists in estimating the error term arising from the approximated mean of an arithmetical function.

If f is an arithmetical function-which we assume here takes real values-and if f behaves in a seemingly erratic manner, it is natural to investigate the summatory function

$$F(x) := \sum_{n \leqq x} f(n), \tag{1.1}$$

in the hope of finding a regular function P satisfying

$$F(x) \sim P(x) \quad (x \to \infty). \tag{1.2}$$

Note:

We shall not attempt here to give an exact definition of what we mean by a "regular function"; it would be long and abstract, and not particularly enlightening nor useful for our present purpose. We agree, however, to acknowledge that all the "logarithmico-exponential functions" of Du Bois-Reymond (see for instance Hardy's tract [9]) are regular, as are the asymptotic expansions in terms of such functions, and their primitives ($\ell_i(x) := \int_2^x dt/\log t$ is an example of both the last two).

If such a function P can be found, $P(x)/x$ may be regarded as a sort of mean for f. Of course P, which we call a principal term for F, is not unique-if it exists at all; its choice must be made by the investigator according to criteria that naturally arise in the course of the investigator's particular investigation.

The next problem consists in estimating the error term implied in (1.2). We write

$$e(x) := F(x) - P(x). \tag{1.3}$$

But of course there is the possibility that, for some regular function $P_1(x)$

$$e(x) \sim P_1(x) \quad (x \to \infty). \tag{1.4}$$

We avoid ending up in such a situation by selecting an appropriate P, in order to obtain what we may call a primary error term in (1.3). This, theoretically, is possible whenever (1.2) exits: otherwise F itself is regular and there is no problem to speak of (we regard regular functions as "known" functions). We may go, in a sense, one step further and define $e(x)$ in such a way that the function

$$\overline{e}(x) := \int_0^x e(t)dt \tag{1.5}$$

be primary (i.e. not asymptotically equivalent to any regular function). In either case we are left with a function having an (often irregular) oscillatory behaviour, which may (partially) be characterized through Ω-type estimates.

But these general remarks are best illustrated by an example. Dirichlet knew in 1849 [5] that the sum-of-divisors function $\sigma(n)$ is on average $\pi^2 n/6$, in the sense that

$$F(x) := \sum_{n \leqq x} \frac{\sigma(n)}{n} \sim \frac{\pi^2}{6} x \quad (x \to \infty). \tag{1.6}$$

If we put, as seems natural,

$$e_0(x) := F(x) - \frac{\pi^2}{6} x , \tag{1.7}$$

e_0 is not a primary function: $e_0(x) \sim -\frac{1}{2} \log x$ as $x \to \infty$. If we take instead

$$e_1(x) := F(x) - \frac{\pi^2}{6} x + \frac{1}{2} \log x , \tag{1.8}$$

e_1 is now primary: indeed

$$e_1(x) = o(\log x) \quad (x \to \infty) \tag{1.9}$$

(see [38] for a better estimate), and [28]

$$e_1(x) = \Omega_\pm(\log \log x). \tag{1.10}$$

However, if we make use of the Perron inversion formula to compute the principal term $P(x)$ of $F(x)$ we obtain

$$P(x) = \sum_{s=0,1} \operatorname{Res} \frac{x^s \zeta(s)\zeta(s+1)}{s} = \frac{\pi^2}{6} x - \frac{1}{2} \log x - \frac{\gamma + \log 2\pi}{2} . \tag{1.11}$$

And although $e_1(x) \neq O(1)$, the constant appearing on the right side of (1.11) is not really an accident and should not be summarily dismissed. Indeed if we define

$$e_2(x) := e_1(x) + \frac{\gamma + \log 2\pi}{2} , \tag{1.12}$$

we have, in the notation of (1.5),

$$\bar{e}_2(x) = \begin{cases} O(x^{1/3}) & (1.13) \\ \Omega_\pm(x^{1/4}) & (1.14) \end{cases}$$

(see Section 5 below for an improvement on (1.13), and [17] for (1.14)), and $e_2(x)$ is the "right" error term, both in the sense that it is on average zero and in the sense explained in (1.5).

But then, one might argue that it is not very natural to let the argument x of F run on all real numbers since it is, after all, the arithmetical function $\sigma(n)$ we are primarily interested in. In view of Pillai and Chowla's [31]

$$\sum_{n \leq x} e_2(n) \sim \frac{\pi^2}{12} x \quad (x \to \infty) \tag{1.15}$$

the "right" error term is thus rather

$$e_3(n) = e_2(n) - \frac{\pi^2}{12}\ , \tag{1.16}$$

since it is on average zero on the integers. Another argument in favour of e_3 is that the distribution function of the values of $e_3(n)$,

$$D(u) := \lim_{x\to\infty} \frac{1}{x}\ \{|n\leq x,\ e_3(n)\leq u\}| \tag{1.17}$$

(it exists and is continuous [6],[30]) is symmetric [30]:

$$D(u) + D(-u) = 1;\ \text{so}\ D(0) = \frac{1}{2}\ . \tag{1.18}$$

However, if we intend to exploit the representation [38],[27]

$$e_2(x) = \sum_{n\leqq\sqrt{x}} \frac{1}{n}\ (\{\frac{x}{n}\} - \frac{1}{2}) + o(1) \tag{1.19}$$

-and this is indeed our intention- e_2 is the error term we must adopt. In general, in the sequel, the error terms associated to divisor functions are all defined like e_2, by an application of the Perron inversion formula.

2. DIRICHLET'S DIVISOR PROBLEM AND THE PILTZ-HARDY-LANDAU CONJECTURE

We put

$$\sigma_0(n) := \sum_{d|n} 1 \quad \text{and} \quad S_0(x) := \sum_{n\leqq x} \sigma_0(n). \tag{2.1}$$

Dirichlet first studied in 1849 [5] the asymptotic behaviour of S_0. If we define the error term by

$$E_0(x) := S_0(x) - x \log x - (2\gamma-1)x - 1/4, \tag{2.2}$$

then it is known, due to Hardy and Ingham (see [35], 12.11), for an up to date account), that

$$E_0(x) = \Omega_\pm(x^{1/4}). \tag{2.3}$$

It is generally believed that the exponent in (2.3) is best possible. According to Landau [21], it was conjectured by Piltz in

1881 and again in 1901 (also see Hardy's [10],[11]) that

$$E_0(x) = O(x^{\frac{1}{4}+\varepsilon}) \quad (\varepsilon > 0). \tag{PHL}$$

An argument in favour of (PHL) is that it is true "on average", in the sense that, with the notation of (1.5),

$$\overline{|E_0|}(x) = O(x^{\frac{5}{4}+\varepsilon}) \quad (\varepsilon > 0). \tag{2.4}$$

This also was proved by Hardy in 1916 [11]. The expression

$$E_0(x) = -2G_{0,1}(x) + O(1), \tag{2.5}$$

where

$$G_{0,1}(x) := \sum_{n \leq \sqrt{x}} (\{\frac{x}{n}\} - \frac{1}{2}), \tag{2.6}$$

is well known [21]. We may thus reformulate

$$\alpha_1(0) = \frac{1}{4}, \tag{PHL}$$

where $\alpha_1(0)$ denotes the smallest α satisfying $G_{0,1}(x) = O(x^{\alpha+\varepsilon})$ for all $\varepsilon > 0$. Much energy has been spent towards proving (PHL), which, however, remains unsettled to date. So far the best published result is due to Kolesnik [19], and Iwaniec and Mozzochi have recently proved [16]

$$\alpha_1(0) \leq 7/22 = 0.318\cdots . \tag{2.7}$$

It is interesting to note that Voronoï [36] had already obtained, in 1904,

$$\alpha_1(0) \leq 1/3 = 0.333\cdots \tag{2.8}$$

and to realize what more than 80 years of hard work on that problem with the help of increasingly sophisticated methods-(2.7) was preceded by seven other successive improvements of (2.8)-have produced.

3. THE CHOWLA-WALUM CONJECTURE

Have we put more generally, for a a real number

$$\sigma_a(n) := \sum_{d|n} d^a \quad \text{and} \quad S_a(x) := \sum \sigma_a(n). \tag{3.1}$$

Berger knew in 1887 [1] the principal term for $S_a(x)$; we denote by $E_a(x)$ the error term (see Cramér's [4]). For $a > -2$ (at least) $E_a(x)$ can be expressed in terms of the functions

$$G_{b,k}(x) := \sum_{n \leqq \sqrt{x}} n^b \psi_k(x/n), \tag{3.2}$$

where $\psi_k(y) := B_k(\{y\})$ and B_k denotes the k-th Bernoulli polynomial. The general formula for $a \geqq 0$ is [15], Theorem 3; for $|a| < 2$ we can then infer from [17], (1.3), [18], (I), [29], Lemma 3, that

$$E_a(x) = -x^a G_{-a,1}(x) - G_{a,1}(x) + O(x^{a/2}) \tag{3.3}$$

(which contains (2.5)).

<u>Remark 3.1</u>. E_{-1} is the e_2 of (1.12).

If we consider now the summatory function

$$S_a^*(x) := \sum_{n \leqq x} (x^a - n^a)\sigma_{-a}(n) \tag{3.4}$$

for $a > 0$, and denote by $R_a(x)$ its error term, then we have by (3.3)

$$R_a(x) = x^a E_{-a}(x) - E_a(x) = O(x^{a/2}). \tag{3.5}$$

<u>Remark 3.2</u>. R_1 is, up to a bounded quantity, the $\bar{e}_2$ of (1.13-14).

<u>Remark 3.3</u>. Estimate (3.5) is actually pretty bad (see below), but the fact that it apparently doesn't depend on the functions $G_{b,1}$ is encouraging: there is an inherent dificulty in estimating $G_{b,1}$ (rather than, say, $G_{b,2}$), that can best be seen through the Fourier expansion

$$\psi_k(t) = \frac{-k!}{(2\pi i)^k} \lim_{N\to\infty} \sum_{m=-N}^{N} \frac{e(mt)}{m^k}. \tag{3.6}$$

Here $e(x) := \exp(2\pi ix)$.

It however took some time before this was discovered and exploited. In 1914 Wigert [39] expanded R_1 in terms of a Bessel function to obtain the rather disappointing-in view of the effort performed- $R_1(x) = O(x^{3/4})$. In 1915 Landau [20] used an expansion, also due to Wigert, of $\overline{R}_1(x)$ in terms of another Bessel function to obtain the more satisfactory (already better than (3.5)) $R_1(x) = O(x^{2/5})$. This was generalized by Wilson [40] in 1922 to $R_a(x)$; here the Perron inversion formula is used. In 1924 Landau again [22] made use of an early version of Van der Corput Lemma, to the effect that (1/2,1/2) is an exponent pair (see Section 4), to prove $R_1(x) = O(x^{1/3})$; however (1/2,1/2), as we shall see, does not in general provide very sharp estimates; indeed, in the same paper, Landau makes use of another method-due to Weyl-for the estimation of exponential sums, to improve upon his own result and obtain $R_1(x) = O(x^{7/23}(\log x)^{12/23}))$.

Then, an imortant step was made in 1936 by Walfisz [37], who discovered the relation

$$R_1(x) = -\frac{1}{2}(G_{0,2}(x) + x^{-1}G_{2,2}(x)) + O(\log x). \tag{3.7}$$

A much more efficient use of estimates for exponential sums could thus-in view of (3.6)-be made, and Peng [24] in 1942 obtained $R_1(x) = O(x^{2/7})$ from the fact that (2/7,4/7) is an exponent pair.

At this point Chowla and Walum [2] in 1963 remarked that the estimate $G_{1,2}(x) = O(x^{3/4})$ could be obtained surprisingly easily. This, with the Piltz-Hardy-Landau conjecture and the dawning realization that the optimal value for the exponent α in $R_1(x) = O(x^{\alpha+\varepsilon})$ $(\varepsilon > 0)$ is probably 1/4, led them to the conjecture

$$\alpha_k(a) \le \frac{a}{2} + \frac{1}{4} \quad (a \ge 0,\ k = 1,2,\cdots) \tag{CW}$$

where $\alpha_k(a)$ denotes the least α for which $G_{a,k}(x) = O(x^{\alpha+\varepsilon})$ is true for each $\varepsilon > 0$ (for the sake of precision, we note that (CW) was originally stated for a an integer).

The relation (3.7) was recently generalized by Kanemitsu [17], who proved that

$$R_a(x) = -\frac{a}{2}\left(x^{a-1}G_{1-a,2}(x) + x^{-1}G_{1+a,2}(x)\right) + O(x^{(a-1)/2}) \quad (3.8)$$

holds for $0 < a < 3$, $a \neq 1$: it follows that in this range estimating R_a reduces to estimating $G_{b,2}$ functions.

Also recently Kanemitsu and Rao [18], (I), proved (CW) in the case where $k \geqq 2$ and $a \geqq 1/2$. The same authors [18], (II), and Srinivasan [34], by proving that -in the sense of (2.4)- (CW) is true on average for $k \geqq 2$ and $a > -1/2$, and Hafner [12], by proving -with Ω- estimates - that (CW) is optimal for $k = 2$, $|a| < 1/2$, naturally led to the more precise conjecture

$$\alpha(a) = \begin{cases} \frac{a}{2} + \frac{1}{4} & (a \geqq -1/2) \\ 0 & (a \leqq -1/2). \end{cases} \quad (S)$$

In the sequel we shall, however, restrict ourselves to the extended (CW) conjecture (S $\leq$) (where we replace in (S) the = sign by $\leq$). For $k \geqq 2$ (S $\leq$) thus remains unsettled if $-1 < a < 1/2$. Nowak [23] by generalizing Peng's result obtained an estimate for $0 \leq a < 1/2$. Recknagel [33] improved it; it follows from his result (via (3.7)) that $R_1(x) = O(x^{109/382})$.

In the following Sections we describe how the method of one-dimensional exponent pairs can be used to improve upon the known O-estimates for the functions $G_{a,k}$ in the ranges $-1 < a < 1/2$, $k \geqq 2$ and $-1 < a \neq 0$, $k = 1$.

As we shall see, the results we obtain depend on the lower border of the set of known exponent pairs; and the lower the border, the better the estimates. What we describe rather extensively in the next Section is done with less details in [29] for Rankin's set of exponent pairs [32], and is used there to obtain in particular

$R_1(x) = O(x^{0.28531755\cdots})$. Due to recent achievements by Huxley and Watt [14] we have now a larger set of exponent pairs than Rankin's to work with; the improvement

$$R_1(x) = O(x^{\frac{37}{130}+\varepsilon}) \tag{3.9}$$

follows as a corollary.

4. ONE DIMENSIONAL EXPONENT PAIRS

We borrow the Graham-Kolesnik [8] simplified version of Phillips' [25] definition of an exponent pair. We set $I := [a,b] \subset [B,2B]$ where $B \geqq 1$ and we deal with the functions f satisfying the following conditions:

(1) f has infinitely many derivatives on I,

(2) there are positive numbers y, s and $d < 1/2$ such that for all integers $r \geqq 0$ and all $x \in I$,

$$|f^{(r+1)}(x) - (-1)^r (s)_r\, yx^{-s-r}| < d(s)_r\, yx^{-s-r}\ , \tag{4.1}$$

where the symbol $(s)_r$ is defined by $(s)_0 = 1$ and $(s)_r = (s)_{r-1}(s+r-1)$ if $r \geqq 1$,

(3) $A := ya^{-s} \geqq 1/2$.

<u>Definition</u>: The pair (κ,λ) of real numbers satisfying $0 \leqq \kappa \leqq 1/2 \leqq \lambda \leqq 1$ is called an exponent pair (e.p. in the sequel) if for all f satisfying conditions (1), (2) and (3) above the estimate

$$\sum_{n\in I} e(f(n)) = O(A^\kappa B^\lambda) \tag{4.2}$$

holds.

Note that (0,1) is trivially an e.p. The following theorem yields other e.p.

<u>Theorem 4.1</u>. (Van der Corput [3], Phillips [25]). *If* (κ,λ) *is an e.p., then*

$$A(\kappa,\lambda) := \left(\frac{\kappa}{2(\kappa+1)} , \frac{1}{2} + \frac{\lambda}{2(\kappa+1)}\right)$$

and, if

$$2\lambda + \kappa \geqq 3/2 , \qquad (4.3)$$

$$B(\kappa,\lambda) := (\lambda - 1/2, \kappa + 1/2)$$

also are e.p. Moreover, the set E *of (all) e.p. is convex in the plane.*

The problem of determining E is very difficult and so far unsolved. Rankin's set S [32] is the subset of E obtained by applying a finite (but arbitrary) number of times Theorem 4.1, starting with the trivial e.p (0,1). S is contained in the triangle Δ of vertices (0,1), (1/2,1/2) and (0,1/2), and contains the triangle Λ of vertices (0,1), (1/2,1/2) and (1/6,2/3). S may be seen as the convex hull conv T of the set T obtainable from (0,1) by applying to it all the finite compositions built with the operators A and BA.

In the sequel, the set of such compositions will be denoted by F. and F_n will be the subset of F consisting of the compositions built with exactly n operators A and BA. Moreover, we shall say that an element D of F_n has length L(D) = n.

A description of recursive constructions of the closure $\bar{T}$ of T, and of S, can be found in [29]. The set $\bar{T}$ enjoys noteworthy properties, as we shall see with the help of the following applications of general results due to Hutchinson [13], 3.1.

Theorem 4.2.

(a) *There is a unique compact set* K *in* Δ *satisfying* $A(K) \cup BA(K) = K$.

(b) K *is the closure of the set of fixed points for the members of* F.

(c) *If we define recursively* $\Delta_0 := \Delta$, $\Delta_{n+1} := A(\Delta_n) \cup BA(\Delta_n)$, *then* $\Delta_n \to K$ *in the Hausdorff metric. Moreover, by* (b), Δ_n *contains* K *for each* n.

Consequences. (1) Since $A(\bar{T}) \cup BA(\bar{T}) = \bar{T}$, it follows by (a) that

$$K = \overline{T} \tag{4.4}$$

(2) A has a Lipschitz constant Lip $A = \sqrt{3}/2$ in Δ. It follows that, uniformly in the set of sequences $\{D_n\}$ $(n = 1,2,\cdots)$ in F satisfying $D_1 = A$ or BA and $D_{n+1} = D_nA$ or D_nBA, the diameter $\delta(D_n(\Delta))$ and the surface area $S(D_n(\Delta))$ both tend to zero as $n \to \infty$. Further, since $D_{n+1}(\Delta) \subset D_n(\Delta)$, $D_n(\Delta)$ converges, to a point in $\overline{T}$ by (c). The Rankin - Graham algorithm RGA I shall mention below produces such sequences.

(3) Note that A transforms a straight line into a straight line. It is easy to see that the 2^n triangles of Δ_n form a chain without loop joining (0,1) and (1/2,1/2), two adjacent triangles having exactly one vertex in common. This implies with (a), (c) and (2) that $\overline{T}$ is the uniform limit of a sequence of curves joining the two points.

(4) Finally we note, for further reference, that if D_n is any sequence in F with $L(D_n) \to \infty$ as $n \to \infty$, then by (c) and (4.4)

$$D_n(\Delta) \to \overline{T} \quad (n \to \infty) \tag{4.5}$$

in the Hausdorff metric.

Whether there exists e.p. outside S remained an open question for many years. The first affirmative answer was given in 1986 by Heath-Brown [35], 6.17-18, who proved that the pair

$$(p,q) = (p,q)_m = ((25(m-2)m^2 \log m)^{-1},\ 1 - (25m^2 \log m)^{-1}) \tag{4.6}$$

is an e.p. for all $m \geqq 3$, and is not in T if $m \geqq 10^6$. (In the latter case the argument given in (6.18) can easily be completed to show that (p,q) is not in S either).

More recently Huxley and Watt [14] proved that for $0 < \varepsilon \leqq 5/56$ the pair

$$h(\varepsilon) := (9/56 + \varepsilon,\ 1/2 + 9/56 + \varepsilon) \tag{4.7}$$

is an e.p. When ε is small enough, $h(\varepsilon)$ is not in S.

By using such new exponent pairs and by applying repeatedly Theorem 4.1, one may now recursively construct a larger subset of E than S. If N is the set of e.p. we start with, we shall denote by $S_1 = S_1(N)$ the set thus obtained. We note that if N is symmetric with respect to the line $L := \{\lambda = \kappa + 1/2\}$, then B need to be applied only to elements of A(E), where condition (4.3) is always satisfied and may thus be ignored. By (4.6) and (4.7) we could start with

$$N := \{h(m^{-1}),\ (p,q)_m,\ B(p,q)_m\ (m \geqq 10^6)\} \tag{4.8}$$

to construct S_1, since condition (4.3) is satisfied by $(p,q)_m$.

However, the contribution of the $(p,q)_m$ to the construction of S_1 is for all practical purposes insignificant. For this reason, and for the sake of simplicity, we briefly describe instead how the set $S_1 = S_1(M)$, where

$$M := \{h(\varepsilon)\ (0 < \varepsilon \leqq 5/56)\} \tag{4.9}$$

can be constructed.

Recursive construction of S_1. Let $\Lambda_0 := \Lambda$ and $h := (9/56, 37/56)$. At the n-th step we apply the elements of F_n to h, call $F_n(h)$ the set of points thus obtained, and construct

$$\Lambda_n := \operatorname{conv}(\Lambda_{n-1} \cup F_1(h)). \tag{4.10}$$

We have

$$S_1 = \overline{\bigcup_{n=0}^{\infty} \Lambda_n}. \tag{4.11}$$

As it was mentioned before, it is the lower borders ∂S and ∂S_1 of S and S_1 we are principally interested in, where we mean by lower border the border from which the segment without endpoints joining (0,1) and (1/2,1/2) has been removed. The reason for this is that in most applications it is the infimum on $R = \partial S$ or ∂S_1 of some function $\theta : \Delta \to \mathbb{R}$ - we denote here and in the sequel this infimum by $I(R,\theta)$ - which produces the optimal result obtainable by the theory of e.p. within the whole set S or S_1.

Recently S.W. Graham [7], clarifying and generalizing a result of Rankin's [32], has given a partial answer to the problem of finding an algorithm that will generate a sequence of e.p. in T approximating the infimum (in $\overline{T}$) of $\theta : \Delta \to \mathbb{R}$ defined by

$$\theta(\kappa,\lambda) := \frac{a\kappa + b\lambda + c}{d\kappa + e\lambda + f}, \tag{4.12}$$

where a,b,c,d,e,f are real constants. This infimum is a value taken by the function θ at a certain $(\bar{\kappa},\bar{\lambda}) \in \overline{T}$, but it is not difficult to see that by the nature of the problem we have in fact

$$I(T,\theta) = I(S,\theta). \tag{4.13}$$

The algorithm (RGA in the sequel) produces a sequence $\{D_n\}$ in F with $L(D_n) \to \infty$ as $n \to \infty$, and such that $\{\theta(D_n(0,1))\}$ converges to $I(S,\theta)$. It has been suggested that this method might be extended to yield $I(S_1,\theta) = I(\partial S_1,\theta)$ instead. But as a converging process this is clearly impossible by (4.5): starting with any e.p. (in fact, starting with any pair in Δ) such an algorithm will invariably produce a sequence of e.p. converging, in the Hausdorff metric, to $\overline{T}$. There is, however, the hope that some term (κ,λ) of the sequence might satisfy $\theta(\kappa,\lambda) < I(S,\theta)$.

In fact, under the assumptions that

$$I(S_1,\theta) < I(S,\theta), \tag{4.14}$$

and that the numerical value of $I(S,\theta)$ is available with an arbitrarily good accuracy, the following procedure will yield $I(S_1,\theta)$ in a finite number of steps.

Brute Force Algorithm (BFA). Compute $D(h)$ for all $D \in F_1$, then for all $D \in F_2$, and so forth. After a finite number of computations we must, by (4.14), find some D_0 for which $I(S,\theta) - \theta(D_0(h)) := \varepsilon > 0$. Now by the continuity of θ in Δ there is $\delta > 0$ such that whenever $d(a,b) < \delta$ for a and b in Δ we have $|\theta(a) - \theta(b)| < \varepsilon$. Finally, by (4.5), there is an N such that if $n \geqq N$ and $D \in F_n$ then $d(D(h),\overline{T}) < \delta$: the desired infimum is $\min\{D(h), L(D) < N\}$.

4A. APPLICATIONS OF THE BFA

Example 4.1. The maximal order of magnitude of $|\zeta(1/2+it)|$.

Let μ denote the smallest ν for which $\zeta(1/2+it) = O(t^{\nu+\varepsilon})$ for every positive ε. It is known that if (κ,λ) is an e.p. and if $\theta_1(\kappa,\lambda) := \kappa+\lambda$, then $\mu \leq \theta_1(\kappa,\lambda)/2 - 1/4$. Thus

$$\mu \leq I(S_1,\theta_1)/2 - 1/4 . \tag{4.15}$$

With the BFA we see that $I(S_1\theta_1) = \theta_1(h) = 23/28$, whence

$$\mu \leq 9/56. \tag{4.16}$$

In fact, the discovery of h by Huxley and Watt is a corollary of their simplified and improved version, in [14], of the proof of (4.16), originally due to Bombieri and Iwaniec.

Example 4.2. Dirichlet's divisor problem and Rankin's constant.

Van der Corput proved [3] that if (κ,λ) is an e.p., then $\alpha_1(0) \leq (\kappa,\lambda)/2(\kappa+1) =: \theta_2(\kappa,\lambda)$. Thus

$$\alpha_1(0) \leq I(S_1,\theta_2) \leq I(S,\theta_2) = I(T,\theta_2) = I(T,\theta_1) - \frac{1}{2} = 0.32902\cdots, \tag{4.17}$$

since by definition every (k,ℓ) in T is $A(\kappa,\lambda)$ for some (κ,λ) also in T (RGA applied to θ_1 yields the numerical value). There is unfortunately no corresponding property in S_1. With BFA we see that

$$\alpha_1(0) \leq I(S_1,\theta_2) = \theta_2(BA^2(h)) = 0.32894\cdots \tag{4.18}$$

(see (2.7)).

Example 4.3. Large values of the zeta function (see [7], 5.II).

Let $t_1 < t_2 < \cdots < t_M$ be points such that $0 \leq t_j < T$, $t_{j+1} - t_j \geq 1$, and $|\zeta(1/2 + it_j)| \geq V$. Define Θ as the infimum of the θ such that if $V \geq T^\theta$, then

$$M = O(T^{1+\varepsilon}V^{-6}).$$

It is known that if (κ,λ) is an e.p., then $\Theta \leq \theta_3(\kappa,\lambda) := \lambda/(4\lambda+2-2\kappa)$. Hence by applying RGA to θ_3 we obtain

$$\Theta \leq I(S,\theta_3) = 0.15274\cdots . \tag{4.20}$$

BFA yields

$$\Theta \leq I(S_1,\theta_3) = \theta_3(BA\ BA(h)) = 0.15269\cdots \ . \tag{4.21}$$

Remark 4.1. I am unable so far to decide whether the BFA will in each case yield a result. More precisely, if θ is as in (4.12) and if $I(R,\theta) = I(\partial_1 R,\theta)$, where R is S or S_1, and where $\partial_1 R$ denotes $\partial R\backslash\{(0,1),(1/2,1/2)\}$, is it true that

$$I(S_1,\theta) < I(S,\theta)\ ? \tag{4.22}$$

To prove that (4.22) is true for all such θ it would clearly be sufficient to show that

$$\partial_1 S_1 \cap \partial_1 S = \emptyset. \tag{4.23}$$

I suspect however that (4.23) might be false. A good candidate for a counterexample appears to be, after some computations, the (κ_0,λ_0) of Rankin [32], p.150, that yields $\theta_1(\kappa_0,\lambda_0) = I(S,\theta_1)$.

5. O-ESTIMATES FOR THE $G_{a,k}$

We now pass to the results announced at the end of Section 3. The proofs are very similar to the arguments of [29], to which we refer the reader.

Define the function $\Gamma : [0,1/2] \to [0,1/2]$ by

$$\partial S_1 =: \{(\kappa,\Gamma(\kappa)),\kappa\in[0,1/2]\}. \tag{5.1}$$

Clearly, Γ is continuous, decreasing and convex. Also define

$$\alpha := I(S_1,\theta_2) = 0.32894\cdots , \tag{5.2}$$

where θ_2 is as in Example 4.2, and

$$\beta := \max\{(\lambda-\kappa)/(\kappa+1),\theta_2(\kappa,\lambda) = \alpha\}. \tag{5.3}$$

For $k = 1$ we have

Theorem 5.1.

$$\alpha_1(a) \leq \begin{cases} \frac{\kappa}{\kappa+1} & \textit{if} \quad a = \frac{\kappa-\Gamma(\kappa)}{\kappa+1}\in[-1,-\beta] \\ a/2+\alpha & \textit{if} \quad a \geq -\beta. \end{cases} \tag{5.4}$$

In view of (3.3), if we denote by $\alpha_E(a)$ the smallest η such that $E_a(x) = O(x^{\eta+\varepsilon})$ is satisfied for every $\varepsilon > 0$, an immediate Corollary of this Theorem is

Corollary 5.1. *Let* $|a| = (\Gamma(\kappa)-\kappa)/(\kappa+1)$. *Then we have*

$$\alpha_E(a) \leqq \begin{cases} a/2 + \alpha & \textit{if} \quad |a| \leqq \beta \\ a/2 + |a|/2 + \kappa/(\kappa+1) & \textit{if} \quad \beta \leqq |a| < 1. \end{cases} \tag{5.5}$$

The proof of Theorem 5.1 uses the material described in Section 4, and (3.6). Owing to the absolute convergence of the latter for $k \geqq 2$ we obtain

Theorem 5.2. *For* $k \geqq 2$ *and* $a = 2\kappa - \Gamma(\kappa) \in [-1,1/2]$, *we have*

$$\alpha_k(a) \leqq \kappa . \tag{5.6}$$

In view of (3.7), (3.8) and [18], (I), this yields, with an obvious notation,

Corollary 5.2. *For* $a = 1 - 2\kappa + \Gamma(\kappa) \in [1/2,2]$, *we have*

$$\alpha_R(a) \leqq a - 1 + \kappa. \tag{5.7}$$

This for $a = 1$ implies (3.9) or, in view of Remark 3.2, an improvement on (1.13). Indeed, $\alpha_R(1) \leqq \kappa|_{\Gamma(\kappa)=2\kappa} \leqq 37/130$, since $BA(9/56,37/56) = (37/130,74/130)$.

Remark 5.1. The statements of Theorems 5.1 and 5.2, and of their respective Corollaries, are those of the results of [29], for which, however, S_1 must be replaced by S in the definitions (5.1), (5.2) and (5.3). Hence the estimates given here are for every a at least as good as those from [29].

Remark 5.2. The curve Γ can be computed arbitrarily well by using the recursive construction of S_1 described in Section 4.

REFERENCES

[1] Berger, A.: Recherches sur les valeurs moyennes dans la théorie des nombres, Nova Acta Reg. Soc. Sc. Ups. Ser. III, pp.130 (1887).

[2] Chowla, S. and Walum, H.: On the divisor problem, Norske Vid. Selsk. Forh., 36, 127-134 (1963).

[3] Corput, J.G. van der: Verschärfung der Abschätzung beim Teilerproblem, Math. Ann., 87, 39-65 (1922).

[4] Cramér, H.: Contributions to the analytic theory of numbers, Fifth congress of Scand. Math. Helsingfors, 266-272 (1922).

[5] Dirichlet, P.J.G.L.: Über die Bestimmung der mittleren Werthe in der Zahlentheorie, Abh. Kön. Preuß Akad., 69-83 (1849), or Werke II, 51-66, Berlin, 1897.

[6] Erdös, P. and Shapiro, H.N.: The existence of a distribution function for an error term related to the Euler function, Canad. J. Math., 7, 63-75 (1955).

[7] Graham, S.W.: An algorithm for computing optimal exponent pairs, J. London Math. Soc., (2) 33, 203-218 (1986).

[8] Graham, S.W. and Kolesnik, G.: One and two dimensional exponential sums, Proc. Oklahoma Number Theory Conf., to appear.

[9] Hardy, G.H.: Orders of infinity. The "Infinitärcalcül" of Paul du Bois-Reymond, Cambridge Tracts in Math., 12 (1910).

[10] Hardy, G.H.: On Dirichlet's divisor problem, Proc. London Math. Soc., (2) 15, 1-25 (1916).

[11] Hardy, G.H.: The average orders of the arithmetical functions P(x) and $\Delta(x)$, Proc. London Math Soc., (2) 15, 192-213 (1916).

[12] Hafner, J.L.: On the average order of a class of arithmetical functions, J. Number Theory, 15, 36-76 (1982).

[13] Hutchinson, J.E.: Fractals and self similarity, Indiana Univ. Math. J., 30, 713-747 (1981).

[14] Huxley, M.N. and Watt, N.: Exponential sums and the Riemann zeta function, preprint.

[15] Ishibashi, M. and Kanemitsu, S.: Fractional part sums and divisor functions I, Number Theory and combinatorics, ed. J. Akiyama et al, World Sci. Publ., 119-183, 1985.

[16] Iwaniec, H. and Mozzochi, C.J.: On the divisor and circle problems, preprint.

[17] Kanemitsu, S.: Omega theorems for divisor functions, Tokyo J. of Math., 7, 399-419 (1984).

[18] Kanemitsu, S. and Rao, R.S.R.C.: On a conjecture of S. Chowla and of S. Chowla and H. Walum I and II, J. Number Theory, 20, 255-261; 103-120 (1985).

[19] Kolesnik, G.: On the method of exponent pairs, Acta Arith., 45, 115-143 (1985).

[20] Landau, E.: Wigert, S. (Stockholm): Sur quelques fonctions arithmétiques, Gött. gel. Anz. 177, 377-414 (1915), or Collected Works 6 (Thales Verlag, ed. P.T. Bateman et al., undated, 198..), 270-307.

[21] Landau, E.: Über Dirichlets Teilerproblem, Gött. Nachr., 13-22 (1920), or Collected Works, 7, 232-251 (198..).

[22] Landau, E.: Über einige zahlentheoretische Funktionen, Gött. Nachr., 116-134 (1924), or Collected Works, 8 (to appear).

[23] Nowak, W.G.: On a problem of S. Chowla and H. Walum, Bull. Number Theory, 7, 1-10 (1983).

[24] Peng, H.Y.: A result in divisor problem, Acad. Sinica Sci. Rec., 1, 69-72 (1942).

[25] Phillips, E.: The zeta function of Riemann; further developments on van der Corput's method, Quart. J. Math., (Oxford), 4, 209-225 (1933).

[26] Pétermann, Y.-F.S.: An Ω-theorem for an error term related to the sum-of-divisors function, Mh. Math., 103, 145-157 (1987).

[27] Pétermann, Y.-F.S.: Existence of all the asymptotic λ-th means for certain arithmetical convolutions, to appear in Tsukuba J. Math.

[28] Pétermann, Y.-F.S.: About a theorem of Paolo Codecà's and Ω-estimates for arithmetical convolutions, preprint.

[29] Pétermann, Y.-F.S.: Divisor problems and exponent pairs, to appear in Arch. Math.

[30] Pétermann, Y.-F.S.: On the distribution of values of an error term related to the Euler function, preprint.

[31] Pillai, S.S. and Chowla, S: On the error terms in some asymptotic formulae in the theory of numbers II, J. Indian Math. Soc., 18, 181-184 (1930).

[32] Rankin, R.A.: Van der Corput's method and the theory of exponent pairs, Quart. J. Math., Oxford (2), 6, 147-153 (1955).

[33] Recknagel, W.: Über eine Vermutung von S. Chowla und H. Walum, Arch. Math., 44, 348-354 (1985).

[34] Srinivasan, S.: A footnote to a conjecture of S. Chowla and H. Walum, Rend. Acad. Naz. Sci. dei XL Mem. Mat., 104, X 39-42 (1986).

[35] Titchmarsh, E.C., The Theory of the Riemann Zeta-function, 2nd ed. revised by D.R. Heath-Brown, Clarendon Press, Oxford, 1986.

[36] Voronoï, G.: Sur une fonction transcendante et ses applications à la sommation de quelques séries, Ann. Sci. Ecole Norm. Sup. 3es., 21, 207-267; 459-533 (1904).

[37] Walfisz, A.: Teilerprobleme, vierte Abhandlung, Ann. Scuola Norm. Sup. Pisa 2as., 5, 289-298 (1936).

[38] Walfisz, A., Weylsche Exponentialsummen in der neueren Zahlentheorie, VEB Deutscher Verlag der Wissenschaften, Berlin, 1963.

[39] Wigert, S.: Sur quelques fonctions arithmétiques, Acta Math., 37, 113-140 (1914).

[40] Wilson, B.M.: An asymptotic relation between the arithmetic sums $\sum_{n\leq x} \sigma_r(n)$ and $x^r\sum_{n\leq x} \sigma_{-r}(n)$, Proc. Cambridge Phil. Soc., 21, 140-149 (1922).

Proc. Prospects
of Math. Sci.
World Sci. Pub.
231-234, 1988

ON G-FUNCTIONS

XU GUANGSHAN
Institute of Mathematics
Academia Sinica
Beijing
People's Republic of China

The main results in this paper are a joint work with K. Väänänen.

In his fundamental paper in 1929, Siegel [3] defined the two classes of analytic functions satisfying linear differential equations, which are called F- and G-functions. An entire function f(z) is called E-function, if it is written of the form

$$f(z) = \sum_{n=0}^{\infty} a_n z^n/n! \ ,$$

where the coefficients a_n belong to an algebraic number field $\mathbb{K}$ and satisfy the conditions:

(i) $\overline{|a_n|} = O(n^{\varepsilon n})$ for all n and for any $\varepsilon > 0$;

(ii) There exists a sequence $\{q_n\}$ of natural numbers such that for all n, $q_n = O(n^{\varepsilon n})$

and for all n and $\ell=1,\cdots,n$, $q_n \cdot a_\ell \in O_{\mathbb{K}}$, where $O_{\mathbb{K}}$ is the ring of integers of $\mathbb{K}$.

Siegel first developed a method for studying the arithmetic properties of the values of E-functions and proved the algebraic independence of $J_0(\xi)$ and $J_0'(\xi)$ for every algebraic number $\xi \neq 0$, where $J_0(z)$ is the Bessel function. He also pointed out that his method could be used to investigate G-functions, and gave some examples of the results that could be obtained. The classical definition of G-function is as follows. The function $g(z)$ is written of the form

$$g(z) = \sum_{n=0}^{\infty} a_n z^n ,$$

where the coefficients $a_n \in \mathbb{K}$ and satisfy the conditions (i) and (ii) with $n^{\varepsilon n}$ replaced by C^n, where C is a constant with $C>1$. This suggestion of Siegel about G-function has been followed recently by Nurmagomedov, Galochkin, Flicker, Väänänen, Matveev and Xu ···, but the results of these papers use the additional Galochkin's condition for the system of linear differential equations satisfied by G-functions except the definition of G-function. In an important paper of Bombieri[1], this condition is replaced by another condition. Then, in a recent paper, Chudnovsky[2], by using ingenions new ideas, succeeded in considering the arithmetic properties of the values of classical G-functions without any restrictive condition. In particular, he gave a lower bound for linear forms in the values of G-functions at certain rational points.

The purpose of this paper is to give some generalizations of Chudnovsky's results to algebraic number fields, both in archimedian and p-adic case. Our proof is based on the ideas of Chudnovsky and on local to global technique used as in the work of Bombieri.

Let $g_1(z),\cdots,g_n(z)$ be a set of G-functions which satisfy the conditions (i), (ii) and the system of differential equations

$$\frac{d}{dz} y_i(z) = \sum_{j=1}^{n} A_{ij}(z) y_j(z), \quad i=1,\cdots,n, \tag{1}$$

where all $A_{ij}(z)$ are rational functions over $\mathbb{K}$. Let $d = [\mathbb{K}:\mathbb{Q}]$. For every place v of $\mathbb{K}$ we normalize the absolute value $|\ |_v$ so that if $v|p$, then $|p|_v = p^{-d_v/d}$, and if $v|\infty$, then $|x|_v = |x|^{d_v/d}$, where $d_v = [\mathbb{K}_v:\mathbb{Q}_v]$ and $|\ |$ denotes the ordinary absolute value in $\mathbb{R}$ or in $\mathbb{C}$. We use α_v to denote 1, if $v|p$, and $d_{v/d}$, if $v|\infty$. We denote a linear form by $L(z) = H_0 + \sum_{i=1}^{n} H_i g_i(z)$, where all H_i are elements of $\mathbb{K}$, not all zero. By L_v we mean a linear form obtained by considering L in the corresponding completion $\mathbb{K}_v$. We use the absolute height $h(x)$ of $x \in \mathbb{K}$ defined by $h(x) = \prod_v \max(1,|x|_v)$. The notation $h(H)$ means $h(H) = \prod_v \max(1, \max_{0 \leq i \leq n} |H_i|_v)$.

Theorem 1. (Väänänen and Xu[4]) *Suppose that G-functions* $g_1(z),\cdots,g_n(x)$ *and 1 are linearly independent over* $\mathbb{K}(z)$. *Let* u *and* ε, $0<u$, $\varepsilon<1$, *be given positive numbers. There exists an effective positive constant* λ, *depending only on* u, *and the functions* $\{g_i(z)\}$, *such that if* $\theta \in \mathbb{K}$ *is non-zero and different from the poles of* $A_{ij}(z)$, *and satisfies* $\log h(\theta)>\lambda$,

$$\log|\theta|_v < \min\{(\frac{u\varepsilon}{(n+1)(n+\varepsilon)} - 1)\log h(\theta), -\alpha_v \log 2c\}.$$

Then we have:

$$\log|L_v(\theta)|_v > -(n+1+\varepsilon)\log h(H)+\log^+ \max_i(|H_i|_v),$$

for all $h(H) \geq C_0$, *where* C_0 *is a positive constant depending on* u, ε, θ *and the system* (1). *We denote* $\log^+ a=\log\max(1,a)$, $a \geq 0$.

Remark. If $\mathbb{K}=\mathbb{Q}$ and $v|\infty$, we obtain the Chudnovsky's result [2] from Theorem 1, but here the constants λ and C_0 can be explicitly given. If $\mathbb{K}=\mathbb{Q}$ and $v|p$, we obtain a p-adic analogue to the above result from Theorem 1.

Further, we prove the algebraic independence of the values of G-functions at algebraic points and give the lower bounds both in archimedian and p-adic case. Let $P \in \mathbb{K}[x_1,\cdots,x_n]$ $P \neq 0$, be a polynomial of degree at most s and height $h(P)$.

Theorem 2. (Väänänen and Xu[5]) *Suppose that G-functions* $g_1(z),\cdots,g_n(z)$ *are algebraically independent over* $\mathbb{K}(z)$. *There then exist positive constants* C_1, τ *depending only on* $\{g_i(z)\}$ *and* n, *such that for any* $\theta \in \mathbb{K}$ *being non-zero and different from the poles of* $A_{ij}(z)$, *and satisfying* $h(\theta) \leq h \geq e^e$,

$$\log h \geq (1+\max(3,s))^{4n}\log\log h,$$

and

$$|\theta|_v < e^{-c_1 s(\log h)^{(4n-1)/4n}(\log\log h)^{1/4n}}.$$

Then we have:

$$|P(g_1(\theta),\cdots,g_n(\theta))|_v > h(P)^{-\tau(\log h)^{1/4}(\log\log h)^{-1/4}}$$

for all $h(P) \geqq H$, *where* H *is an effective positive constant depending only on* $\{g_i(z)\}$, n, s, θ, v *and the system* (1).

REFERENCES

[1] Bombieri, E.: On G-functions, Recent Progress in Analytic Number Theory, Vol.II, H. Halberstam and C. Hooley (eds.), Academic Press, London, 1-67 (1981).

[2] Chudnovsky, G.V.: On applications of diophantine approximations, Proc. Nat. Acad. Sci. USA, 81, 7261-7265 (1984).

[3] Siegel, C.L.: Uber einige Anwendungen diophantischer Approximationen, Abh. Preuss. Akad. Wiss., Phys.-Math. Kl. 1, (1929).

[4] Väänänen, K. and Xu Guangshan: On linear forms of G-functions, Acta Arith., (1988) (to appear).

[5] Väänänen, K. and Xu Guangshan: On the arithmetic properties of the values of G-functions, preprint (1986).

Proc. Prospects
of Math. Sci.
World Sci. Pub.
235-266, 1988

ALGEBRAIC VALUES OF FUNCTIONS ON THE UNIT DISK

ISAO WAKABAYASHI

Department of Mathematics
Tokyo University of Agriculture and Technology
Fuchu, Tokyo 183
Japan

ABSTRACT

The Schneider-Lang theorem was generalized by Gramain-Mignotte-Waldschmidt to functions on the unit disk. We improve their results. Namely, we give better estimates from above for the number of points where algebraically independent functions on the unit disk take simultaneously algebraic values or algebraic derivatives. This is achieved by estimating more precisely certain terms in Jensen's formula.

0. INTRODUCTION

The Schneider-Lang theorem gives estimates from above for the number of points where algebraically independent meromorphic functions of moderate growth on the complex plane take simultaneously algebraic values or algebraic derivatives. This theorem was generalized by Gramain-Mignotte-Waldschmidt [1] to functions on the unit disk (see Theorem 1, Corollaries 1 and 2). The aim of this article is to improve their results. For Schneider's method, namely for the case concerning algebraic values of functions, compare Theorem 1 with Theorem 2, Corollary 1 with Corollary 3, and Corollary 2 with Corollary 4, and see Remark 1.1. For Gel'fond's method, namely for the case concerning algebraic derivatives, see Theorem 3, Corollary 5 and Remark 1.3.

The idea of the proof is the following. As usual, we use Jensen's formula (see Lemma 2.1), as an analytic tool, to estimate from above certain values (or derivatives) of an auxiliary function. However we give more precise estimates for certain terms in Jensen's formula (see Lemma 2.2), which are essential for our proof. For an auxiliary function we take the usual one. But for a set at which we require the auxiliary function to vanish, we take a special set such that it contains a relatively large number of points, and such that those points are sufficiently close to each other (see Lemmas 3.3 and 4.5). Then, by the precise estimate mentioned above, we see that the terms in Jensen's formula for those close points are relatively big negative numbers. Therefore, we can obtain useful upper bounds of values (or derivatives) of the auxiliary function, even if it has a smaller number of zeros. This means that, to obtain a contradiction, it is sufficient that the given functions take algebraic values (or derivatives) at a smaller number of points. Thus we obtain the results.

1. STATEMENT OF RESULTS

We shall recall the generalization of the Schneider-Lang theorem by Gramain-Mignotte-Waldschmidt [1], and give some Corollaries. Next, we shall state our results. We shall treat both Schneider's method and Gel'fond's method.

Notation. For an algebraic number field K over $\mathbb{Q}$, we denote by $[K:\mathbb{Q}]$ its degree over $\mathbb{Q}$, and by I_K the ring of algebraic integers in K. For an algebraic number α, we denote by $\overline{|\alpha|}$ the maximum of the absolute values of its conjugates, and by $\operatorname{den} \alpha$ the (smallest) denominator of α. For a function $f(z)$, let $f^{(k)}(z)$ denote its k-th derivative, and let $|f|_R = \max_{|z| \leq R} |f(z)|$. Let $D = \{z \in \mathbb{C} \mid |z| < 1\}$ be the unit disk. For a positive number $N > 1$, let $D_N = \{z \in \mathbb{C} \mid |z| < 1 - N^{-1}\}$ be the disk of radius $1 - N^{-1}$.

The result of Gramain-Mignotte-Waldschmidt is formulated in a very general form. But here we formulate it, for the sake of simplicity, in a slightly restricted form which, we think, is not a strong restriction. However, we note that their original result covers also the case of the whole complex plane.

Theorem 1 (Gramain-Mignotte-Waldschmidt [1]). *Let* $f_1(z),\cdots,f_h(z)$ $(h \geq 2)$ *be algebraically independent meromorphic functions on the unit disk* D. *Let* $\{\Gamma_N\}_{N=1}^{\infty}$ *be a sequence of finite subsets of* D *with* $\Gamma_N \subset D_N$. *Suppose there exist holomorphic functions* g_i, h_i *on* D, $1 \leq i \leq h$, *such that* $f_i = h_i/g_i$ *and* $g_i \neq 0$ *on* Γ_N *for all* N. *Suppose further there exist positive constants* κ, C, C', *and non-negative constants* μ, ρ_i *and* ρ_i', $1 \leq i \leq h$, *such that the following conditions are satisfied for* $i = 1,\cdots,h$ *and all sufficiently large natural numbers* N:

(1) $\operatorname{card} \Gamma_N \geq CN^{\kappa}$;

(2) $|g_i|_{1-1/N}$, $|h_i|_{1-1/N} \leq \exp(C'N^{\rho_i})$, *and* $|g_i(w)| \geq \exp(-C'N^{\rho_i})$ *for any* $w \in \Gamma_N$;

(3) *for every point* $w \in \Gamma_N$,

(i) *there exists a number field* K_w *such that* $f_i(w) \in K_w$ *and* $[K_w:\mathbb{Q}] \leq C'N^{\mu}$;

(ii) $|\overline{f_i(w)}| \leq \exp(C'N^{\rho_i'})$;

(iii) *there exist natural numbers* δ_{iw} *such that* $\delta_{iw} \cdot f_i(w) \in I_{K_w}$ *and* $\delta_{iw} \leq \exp(C'N^{\rho_i'})$.

Then

$$\kappa \leq \max\left\{\frac{2h}{h-1} + \frac{\mu}{h-1} + \frac{1}{h-1}\sum_{i=1}^{h} \max\{\rho_i, \mu+\rho_i'\},\ 2 + \max_{1\leq i\leq h}\{\rho_i, \mu+\rho_i'\}\right\}.$$

Proof. We indicate briefly how Theorem 1 is deduced from Theorem 2.1 of [1]. It is enough to choose the numbers and functions appearing there as follows: $k = h$, $\mu(N) = CN^{\kappa}$, $\theta_i(x) = (1-x)^{-\rho_i}$, $\psi_i(N) = N^{\rho_i'}$, $d(N) = C'N^{\mu}$, $r(N) = 1 - N^{-1}$, $R(N) = 1 - N^{-1-\varepsilon}$ with a sufficiently small positive constant ε. Suppose κ is greater than the right-hand side of the inequality in the conclusion of Theorem 1. Then all the assumptions of Theorem 2.1 of [1] are satisfied, and it implies that $f_1,\cdots,f_h$ are algebraically dependent, which is a contradiction. Hence we get the result. Note that in Theorem 2.1 of [1], f_i are assumed to be holomorphic on D. However its generalization to meromorphic functions is easy, as is mentioned there also.

Corollary 1. *If* $h = 2$, *then under the same assumptions as Theorem* 1,

$$\kappa \leqq 4 + \mu + \sum_{i=1}^{2} \max\{\rho_i, \mu + \rho_i'\}.$$

This Corollary for the case $\rho_i' = \rho_i = h$, $\mu = 0$ is analogous to Corollary 2.2 of [1]. We give another simpler Corollary.

Corollary 2. *Let* $f_1(z)$, $f_2(z)$ *be algebraically independent holomorphic functions on the unit disk* D. *Let* $\{\Gamma_N\}_{N=1}^{\infty}$ *be a sequence of finite subsets of* D *with* $\Gamma_N \subset D_N$. *Let* K *be a number field of finite degree. Suppose there exist positive constants* κ, C, C' *and* C" *such that the following conditions are satisfied for* $i = 1, 2$, *and all sufficiently large natural numbers* N:

(1) $\operatorname{card} \Gamma_N \geqq CN^{\kappa}$;

(2) $|f_i|_{1-1/N} \leqq C'N^{C''}$;

(3) *for every point* $w \in \Gamma_N$,

(i) $f_i(w) \in K$;

(ii) $|\overline{f_i(w)}| \leqq C'N^{C''}$;

(iii) *there exists a natural number* δ_w *such that* $\delta_w \cdot f_i(w) \in I_K$ *and* $\delta_w \leqq C'N^{C''}$.

Then

$$\kappa \leqq 4.$$

Proof. It is enough to put in Theorem 1, $h = 2$, $h_i = f_i$, $g_i \equiv 1$, $\mu = 0$, and $\rho_i = \rho_i' = \varepsilon$ with an arbitrary small positive constant ε.

Now, our first generalization of the Schneider-Lang theorem (Schneider's method) to functions on the unit disk is as follows.

Theorem 2 (Schneider's method). *Under the same assumptions as Theorem* 1,

$$\kappa \leqq \max\left\{ 2 + \frac{\mu}{h-1} + \frac{1}{h-1}\sum_{i=1}^{h} \max\{\rho_i, \mu+\rho_i'\},\ 2 + \max_{1\leqq i\leqq h}\{\rho_i, \mu+\rho_i'\}\right\}.$$

Corollary 3. *If* $h = 2$, *then under the same assumptions as Theorem* 1,

$$\kappa \leqq 2 + \mu + \sum_{i=1}^{2} \max \{\rho_i, \mu + \rho_i'\}.$$

Corollary 4. *Under the same assumptions as Corollary* 2,

$$\kappa \leqq 2.$$

Remark 1.1. The number $\frac{2h}{h-1}$ in Theorem 1 is replaced by 2 in Theorem 2. The number 4 in Corollaries 1 and 2 is replaced by 2 in Corollaries 3 and 4. This is our improvement.

Now let us state our second generalization of the Schneider-Lang theorem (Gel'fond's method) to functions on the unit disk.

Theorem 3 (Gel'fond's method). *Let* $f_1(z)$, $f_2(z)$ *be algebraically independent meromorphic functions on the unit disk* D. *Let* $\{\Gamma_N\}_{N=1}^{\infty}$ *be a sequence of finite subsets of* D *with* $\Gamma_N \subset D_N$. *Suppose there exist holomorphic functions* g_i, h_i *on* D, $i = 1, 2$, *such that* $f_i = h_i/g_i$ *and* $g_i \neq 0$ *on* Γ_N *for all* N. *Suppose further there exist positive constants* κ, C, C', *and nonnegative constants* μ, ρ, ρ', ρ'' *such that the following conditions are satisfied for* $i = 1, 2$, *and all sufficiently large natural numbers* N:

(1) $\operatorname{card} \Gamma_N \geqq CN^{\kappa}$;

(2) $|g_i|_{1-1/N}$, $|h_i|_{1-1/N} \leqq \exp(C'N^{\rho})$, *and* $|g_i(w)| \geqq \exp(-C'N^{\rho})$ *for any* $w \subset \Gamma_N$;

(3) *for every point* $w \in \Gamma_N$, *there exist a number field* K_w, *positive integers* δ_{iw}, *and nonnegative integers* δ_{iw}' *such that for* $k = 0,1,2,\cdots$,

(i) $f_i^{(k)}(w) \in K_w$ *and* $[K_w:\mathbf{Q}] \leqq C'N^{\mu}$;

(ii) $\overline{\left|f_i^{(k)}(w)\right|} \leqq e^{C'N^{\rho'}(k+1)}\, k^{C'N^{\rho''}k}$;

(iii) $\delta_{iw}^{k+1}\, (k!)^{\delta_{iw}'}\, f_i^{(k)}(w) \in I_{K_w}$, $\delta_{iw} \leqq \exp(C'N^{\rho'})$, *and* $\delta_{iw}' \leqq C'N^{\rho''}$.

Then

$$\kappa \leqq 1 + \mu + \max \{\rho', \rho''\}.$$

Remark 1.2. The number ρ does not appear in the conclusion.

Under stronger assumptions, the above result implies the following.

Corollary 5. *We replace the assumption* (3) *of Corollary* 2 *by the following assumption* (3)':

(3)' *for every point* $w \in \Gamma_N$, *there exist a positive integer* δ_w *and a nonnegative integer* δ'_w *such that for* $k = 0,1,2,\cdots$,

$$\text{(i)} \quad f_i^{(k)}(w) \in K;$$

$$\text{(ii)} \quad \left|\overline{f_i^{(k)}(w)}\right| \leq C'N^{C''(k+1)}\,k^{C'k};$$

$$\text{(iii)} \quad \delta_w^{k+1}\,(k!)^{\delta'_w}\,f_i^{(k)}(w) \in I_K, \quad \delta_w \leq C'N^{C''}, \quad \text{and} \quad \delta'_w \leq C'.$$

Then

$$\kappa \leq 1.$$

Proof. It is enough to put in Theorem 3, $\mu = 0$, $\rho' = \varepsilon$ and $\rho'' = 0$ with an arbitrary small positive constant ε.

Remark 1.3. By the method of [1], one can prove that under the same assumptions as Corollary 5, $\kappa \leq 2$. So, for Gel'fond's method also, our result is better.

2. JENSEN'S FORMULA

We recall Jensen's formula.

Lemma 2.1 (Jensen's formula). *Let* $f(z)$ *be a holomorphic function on a closed disk* $\{|z| \leq R\}$. *Let* $w = re^{i\theta}$ *with* $r < R$. *Suppose* f *has the Taylor expansion at* w,

$$f(z) = \alpha_n(z - w)^n + \alpha_{n+1}(z - w)^{n+1} + \cdots \quad (\alpha_n \neq 0),$$

and let $\{w_\nu\}$ *be the zeros of* f *in* $\{|z| \leq R\} - \{w\}$, *where* w_ν *are counted as many times as indicated by their orders. Then*

$$\log |\alpha_n| = \frac{1}{2\pi}\int_0^{2\pi} \log |f(Re^{i\phi})| \frac{R^2 - r^2}{R^2 - 2Rr\cos(\theta-\phi) + r^2}\, d\phi$$
$$+ \sum_{w_\nu} \log \left| \frac{R(w - w_\nu)}{R^2 - \overline{w}_\nu w} \right| - n \log \frac{R^2 - r^2}{R}.$$

We shall use this formula to estimate from above the values or derivatives of auxiliary functions at certain points. It is essential in our proof to give a precise estimate of the terms in the sum (over w_ν). To do it, we use the Poincaré metric of the unit disk.

Let z_1 and z_2 be two points in the unit disk D, and let $d(z_1,z_2)$ denote the Poincaré distance between z_1 and z_2 measured by the Poincaré metric of D, $ds = \dfrac{|dz|}{1 - |z|^2}$.

Let $R > 0$, and $w \in \{|z| < R\}$. Put

$$z' = \phi_{R,w}(z) = \frac{R(z - w)}{R^2 - \overline{w}z}.$$

This defines a one to one holomorphic mapping of $\{|z| < R\}$ onto D, sending w to 0. The following lemma gives an estimate of the terms appearing in Jensen's formula.

Lemma 2.2. *Let* $0 < r < R < 1$, *and let* z *and* w *be two points with* $|z|, |w| \leq r$. *Then we have* (i) *and* (ii).

(i) $d(0,\phi_{R,w}(z)) \leq \dfrac{R(1 - r^2)}{R^2 - r^2}\, d(z,w)$.

(ii) *If* $z \neq w$, *then*

$$\log |\phi_{R,w}(z)| < - e^{-2\{R(1-r^2)/(R^2-r^2)\}d(z,w)}.$$

Further, as a special case of (ii), *we obtain* (iii).

(iii) *Let* ε *be a positive number. Then there exists a natural number* $N_0(\varepsilon)$ *such that for any natural number* $N \geq N_0(\varepsilon)$, $R = 1 - N^{-1-\varepsilon}$, *and* $z, w \in \overline{D}_N$ (i.e., $|w|, |z| \leq 1 - N^{-1}$) *with* $z \neq w$, *we have*

$$\log |\phi_{R,w}(z)| < - e^{-(2+\varepsilon)d(z,w)}. \tag{2.1}$$

Proof. (i) Let $ds' = \dfrac{|dz'|}{1 - |z'|^2}$ be the Poincaré metric of the unit disk $\{|z'| < 1\}$. Then, by a direct calculation, we see that by the mapping $z' = \phi_{R,w}(z)$, the Poincaré metric changes as

$$ds' = \frac{|dz'|}{1-|z'|^2} = \frac{R\,|dz|}{R^2-|z|^2} = \frac{R(1-|z|^2)}{R^2-|z|^2} \cdot \frac{|dz|}{1-|z|^2} = \frac{R(1-|z|^2)}{R^2-|z|^2}\,ds.$$

If $|z| \leqq r$, then by calculation of the maximum of the last term, we obtain

$$ds' \leqq \frac{R(1 - r^2)}{R^2 - r^2}\,ds.$$

Therefore, the Poincaré distance between two points in $\{|z| \leqq r\}$ increases by the mapping $z' = \phi_{R,w}(z)$ at most $R(1 - r^2)/(R^2 - r^2)$ times, which proves (i).

(ii) Put $a = |\phi_{R,w}(z)|$. Then we have

$$d(0,\phi_{R,w}(z)) = d(0,a) = \int_0^a \frac{dr}{1 - r^2} = \frac{1}{2}\log\frac{1 + a}{1 - a}\,.$$

So $a = (e^{2d(0,a)} - 1)/(e^{2d(0,a)} + 1)$, and

$$\log a = \log(1 - (1 - a)) < -(1 - a) = -\frac{2}{e^{2d(0,a)} + 1} < -\,e^{-2d(0,a)}.$$

From (i) this implies (ii).

(iii) Let $r = 1 - N^{-1}$. Then

$$\lim_{N\to\infty} \frac{R(1 - r^2)}{R^2 - r^2} = \lim_{N\to\infty} \frac{(2 - N^{-1})(1 - N^{-1-\varepsilon})}{2 - 2N^{-\varepsilon} - N^{-1} + N^{-1-2\varepsilon}} = 1.$$

Hence (ii) implies (iii), as desired.

Remark 2.1. This Lemma asserts that if the Poincaré distance between two pionts z and w is small, then $\log|\phi_{R,w}(z)|$ is a negative number relatively far from 0. For example, if $d(z,w) \leqq c$, then we obtain an estimate of the form $\log|\phi_{R,w}(z)| < -c' < 0$. This kind of estimate plays an essential role in what follows.

Remark 2.2. If we assume only that two points z and w are included in $\bar{D}_N$, then $d(z,w) \leq \log(2N-1)$, and the equality can be attained. Hence, in comparison with the case of Remark 2.1, we can only get by (iii) a weaker estimate of the form $\log|\phi_{R,w}(z)| < -e^{-(2+\varepsilon)\log(2N-1)} < -c''N^{-2-\varepsilon}$. Therefore, in order to obtain useful upper bounds of values or derivatives of functions by Jensen's formula, one would need a large number of zeros. Our improvement of the results is achieved by constructing auxiliary functions so that we can use estimates as in Remark 2.1.

3. PROOF OF THEOREM 2

We assume that all the conditions of Theorem 2 (see Theorem 1) are satisfied. We assume further that the conclusion of Theorem 2 is false, i.e.,

$$\kappa > 2 + \max\{\frac{\mu}{h-1} + \frac{1}{h-1}\sum_{i=1}^{h}\max\{\rho_i,\mu+\rho_i'\},\ \max_{1\leqq i\leqq h}\{\rho_i,\mu+\rho_i'\}\}, \quad (3.1)$$

and we shall show that this implies a contradiction. We may suppose μ, ρ_i and ρ_i' are positive.

We divide the proof into several steps.

Step 1. *Preliminaries.* We first introduce some quantities. Put

$$\rho = \max\{\frac{\mu}{h-1} + \frac{1}{h-1}\sum_{i=1}^{h}\max\{\rho_i,\mu+\rho_i'\},\ \max_{1\leqq i\leqq h}\{\rho_i,\mu+\rho_i'\}\}. \quad (3.2)$$

Then the assumption (3.1) is written briefly as

$$\kappa > 2 + \rho.$$

In the following, we assume that the index i varies from 1 to h without mentioning it.

Let ε be any positive number such that

$$\kappa > 2 + \rho + 4(1 + \max_i \rho_i)\varepsilon. \quad (3.3)$$

Further, let m be any natural number such that

$$\frac{1}{2^m} < \varepsilon. \tag{3.4}$$

Put

$$\alpha_i = \rho - \max\{\rho_i, \mu+\rho_i'\} + 2(1 + \max_i \rho_i)\varepsilon, \tag{3.5}$$

and

$$\beta = \rho + 3(1 + \max_i \rho_i)\varepsilon. \tag{3.6}$$

These quantities satisfy the following relations which will be used later.

Lemma 3.1. *We have for* $i = 1,2,\cdots,h$,

$$\alpha_i > 0, \tag{3.7}$$

$$\kappa > 2 + \beta + (1 + \max_i \rho_i)\varepsilon, \tag{3.8}$$

$$\kappa - 1 > \beta, \tag{3.9}$$

$$\sum_i \alpha_i > \mu + \beta, \tag{3.10}$$

$$\beta > \mu + \alpha_i + \rho_i', \tag{3.11}$$

$$\beta > \alpha_i + (1 + \varepsilon)\rho_i, \tag{3.12}$$

$$\kappa - 2 - 2\varepsilon > \mu + \alpha_i + \rho_i', \tag{3.13}$$

$$\kappa - 2 - 2\varepsilon > \alpha_i + (1 + \varepsilon)\rho_i. \tag{3.14}$$

Proof. Clearly (3.7) follows from (3.2) and (3.5). (3.8) follows from (3.3) and (3.6). (3.9) follows from (3.8). By (3.2) and the assumption that $h \geqq 2$, we obtain

$$\begin{aligned}\sum_i \alpha_i - \mu - \beta &= (h - 1)\rho - \sum_i \max\{\rho_i, \mu+\rho_i'\} - \mu + (2h - 3)(1 + \max_i \rho_i)\varepsilon \\ &\geqq (\mu + \sum_i \max\{\rho_i, \mu+\rho_i'\}) - \sum_i \max\{\rho_i, \mu+\rho_i'\} - \mu \\ &\quad + (2h - 3)(1 + \max_i \rho_i)\varepsilon \\ &> 0,\end{aligned}$$

so (3.10) holds. Since

$$\begin{aligned}\beta - \mu - \alpha_i - \rho_i' &= \max\{\rho_i, \mu+\rho_i'\} - \mu - \rho_i' + (1 + \max_i \rho_i)\varepsilon \\ &> \varepsilon > 0,\end{aligned} \tag{3.15}$$

(3.11) holds. Since

$$\beta - \alpha_i - (1+\varepsilon)\rho_i = \max\{\rho_i, \mu+\rho_i'\} + (1 + \max_i \rho_i)\varepsilon - (1+\varepsilon)\rho_i$$

$$\geqq \varepsilon > 0, \tag{3.16}$$

(3.12) holds. Since $\kappa - 2 > \beta + \varepsilon$ by (3.8), we obtain (3.13) by (3.15), and (3.14) by (3.16). Thus the proof is complete.

In the following, we shall denote by c_1, $c_2, \cdots$, or c_1', $c_2', \cdots$, positive constants depending only on the quantities so far introduced (and not depending on the following number L). Let L be a sufficiently large integer. We assume that $L^{1/2^m}$ is an integer and at the same time L is a power of 2, namely we assume that L has the form

$$L = 2^{\ell' 2^m} = 2^{\ell} \quad (\ell = \ell' 2^m) \quad \text{with} \quad \ell' \gg 1. \tag{3.17}$$

This is only to avoid taking the integer part of certain numbers, denoted by $[\cdot]$, and to simplify the description. In the following calculation of inequalities, we shall always use the assumption that L is sufficiently large.

For an auxiliary function we shall take the usual one. But for a set at which we require the auxiliary function to vanish, we shall take a special subset S of Γ_L such that card S is relatively large, and such that the points of S are sufficiently close to each other. In fact, we need more, namely we need to construct a suitable sequence $S, S_0, S_1, \cdots, S_m$ of subsets of Γ_L, as is given by Lemma 3.3 below.

Step 2. *Choice of subsets* $S, S_0, S_1, \cdots, S_m$ *of* Γ_L. Our principle is the following. We divide the disk D_L into several parts, and choose one of them which contains the largest number of points of Γ_L. Then we divide this part again into several smaller parts, and choose one of them which contains the largest number of points of Γ_L. We continue this process certain times. Then we shall find a small part of D_L containing relatively many points of Γ_L. We shall take these points finally obtained as the subset S

mentioned above. We shall do this precisely in the following. First we prepare a geometrical lemma. We shall often say simply "diameter" or "area" instead of saying "Poincaré diameter" or "Poincaré area".

Lemma 3.2. *Let* $0 < t \leq 1/2$. *Let* A_t *be a sector of center* 0, (*Euclidian*) *radius* $1 - 1/L$ *and angle* $2\pi/L^t$. *Then*

$$(\textit{Poincaré})\ \textit{diameter of}\ A_t < (1 - t) \log L + c_1.$$

Proof. Let o denote the center of the sector ($o \leftrightarrow z = 0$). Let op be one of the two sides of A_t starting from o. Let q be the point on op such that the Euclidian distance between o and q is $1 - 1/L^t$. Further, let q' be the point on the other side such that the Euclidian distance between o and q' is $1 - 1/L^t$. Then the Poincaré distance between o and q, d(o,q), is given by

$$d(o,q) = \int_0^{1-1/L^t} \frac{dr}{1-r^2} = \frac{1}{2}\{t \log L + \log (2 - 1/L^t)\}.$$

Similarly,

$$d(q,p) = \int_{1-1/L^t}^{1-1/L} \frac{dr}{1-r^2} = \frac{1}{2}\{(1 - t) \log L + \log (\frac{2-1/L}{2-1/L^t})\}.$$

Further, the Poincaré length of the arc qq' is given by

$$\text{length of } qq' = 2\pi(1 - 1/L^t)/(2 - 1/L^t).$$

Hence the Poincaré distance between q and any point of A_t is at most $\frac{1}{2}(1 - t) \log L + c_1'$ since $t \leq 1 - t$ by assumption, which implies the result.

Lemma 3.3. *There exists an increasing sequence* $S \subset S_0 \subset S_1 \subset \cdots \subset S_m$ *of subsets of* Γ_L *satisfying the following properties*:

(i) *diameter of* $S_j < (1 - 1/2^j) \log L + c_1$ *for* $j = 1,2,\cdots,m$; (3.18)

(ii) card $S_j > CL^{\kappa-1/2^j} - 2$ *for* $j = 1,2,\cdots,m$; (3.19)

(iii) *diameter of* $S_0 < c_2$; (3.20)

(iv) card $S_0 > c_3L^{\kappa-1}$; (3.21)

(v) *diameter of* $S < c_2$;

(vi) card $S = [L^\beta]$. (3.22)

Proof. In fact, we shall show that there exist an increasing sequence $A_1 \subset A_2 \subset \cdots \subset A_m$ of sectors of D_L and a subregion A_0 of A_1 such that

(i)' angle of $A_j = 2\pi L^{-1/2^j}$ for $j = 1,2,\cdots,m$;

(ii)' $S_j = A_j \cap \Gamma_L$ satisfies (ii) for $j = 1,2,\cdots,m$;

(iii)' diameter of $A_0 < c_2$;

(iv)' $S_0 = A_0 \cap \Gamma_L$ satisfies (iv),

and a certain subset S of S_0 satisfying (vi). Note that if this is done, then automatically (i) is satisfied by (i)', Lemma 3.2 and the relation $S_j \subset A_j$, and also (iii) and (v) are satisfied by (iii)' and the relation $S \subset S_0 \subset A_0$.

(a) *Choice of* A_j *for* $j = 1,2,\cdots,m$. We shall choose A_j satisfying (i)' and (ii)'.

First we choose A_m. To this aim, we divide the disk D_L into $L^{1/2^m}$ sectors of the same angle, i.e., of angle $2\pi/L^{1/2^m}$. (We assumed for simplicity that $L^{1/2^m}$ is an integer.) Since card $\Gamma_L \geqq CL^\kappa$ by (1) of Theorem 1, one of these sectors contains at least $[CL^\kappa/L^{1/2^m}]$ points of Γ_L. Let A_m denote such a sector. Then clearly A_m satisfies (i)'. Also A_m satisfies (ii)' since card $A_m \cap \Gamma_L \geqq [CL^{\kappa-1/2^m}]$.

Next, let $1 < j \leqq m$, and suppose we have chosen sectors $A_j, A_{j+1}, \cdots, A_m$ with $A_j \subset A_{j+1} \subset \cdots \subset A_m$, such that they satisfy (i)' and (ii)'. We divide A_j into $L^{1/2^j}$ smaller sectors of the same angle. Then, since card $A_j \cap \Gamma_L > CL^{\kappa-1/2^j} - 2$ from the assumption, one of these sectors contains at least $[(CL^{\kappa-1/2^j} - 2)/L^{1/2^j}]$ points of Γ_L. Let A_{j-1} denote such a sector. Then angle of $A_{j-1} = 2\pi L^{-1/2^j}/L^{1/2^j} = 2\pi L^{-1/2^{j-1}}$. So A_{j-1} satisfies (i)'. Also A_{j-1}

satisfies (ii)' since card $A_{j-1} \cap \Gamma_L > CL^{\kappa-1/2^{j-1}} - 2$. Therefore by induction we obtain A_j, for $j = 1,2,\cdots,m$, satisfying (i)' and (ii)'.

(b) *Choice of* A_0. The (Poincaré) area of A_1 is

$$\int_0^{2\pi/L^{1/2}} \int_0^{1-1/L} \frac{r\,dr\,d\theta}{(1-r^2)^2} = \pi\left(\frac{L^{1/2}}{2-1/L} - L^{-1/2}\right) < \pi L^{1/2}.$$

So, noting also that the shape of sector is not so complicated, we can cover A_1 by at most $c_2' L^{1/2}$ subregions, each having bounded area and bounded diameter, say smaller than c_2 (see Remark 3.1 below for an example of such a covering). Then, since card $A_1 \cap \Gamma_L > CL^{\kappa-1/2} - 2$ by (ii)', one of these subregions, say A_0, contains at least $[(CL^{\kappa-1/2} - 2)/(c_2' L^{1/2})]$ points of Γ_L. So, for $c_3 < C/c_2'$, A_0 satisfies (iv)'. Also A_0 satisfies (iii)' clearly.

(c) *Choice of* S. By (3.9), we have $\kappa - 1 > \beta$. So, by (iv)', S_0 contains more than $[L^{\beta}]$ points. We choose arbitrarily $[L^{\beta}]$ points of S_0, and define S to be the set of these points. Then (vi) holds. Thus the proof of Lemma 3.3 is complete.

Remark 3.1. There are many coverings of A_1 having the properties mentioned in (b), and we can use any one. Moreover, the boundedness of the areas of subregions constituting the covering was not necessary in fact for our purpose. In order to complete the proof, we give here an example of covering of A_1 having the properties mentioned in (b).

For simplicity we may suppose

$$A_1 = \{z \in D_L \mid 0 \leq \arg z \leq 2\pi/L^{1/2}\}.$$

Put

$$B_i = \{z \in A_1 \mid 1 - \frac{1}{2^{i-1}} \leq |z| \leq 1 - \frac{1}{2^i}\},$$

for $i = 1,2,\cdots,\ell/2$, and put

$$B_{ij} = \{z \in D \mid 1 - \frac{1}{2^{i-1}} \leq |z| \leq 1 - \frac{1}{2^i},\ \frac{2\pi(j-1)}{2^i} \leq \arg z \leq \frac{2\pi j}{2^i}\},$$

for $i = \ell/2 + 1, \cdots, \ell$ and $j = 1, 2, \cdots, 2^{i-\ell/2}$, where ℓ is given by (3.17). Then A_1 is covered by these B_i and B_{ij}. The number of these (closed) subregions is

$$\frac{\ell}{2} + 2 \cdot 2^{\ell/2} - 2 = \frac{\log L}{2 \log 2} + 2L^{1/2} - 2 < 3L^{1/2} \quad (\text{for } L >> 1).$$

We can easily calculate the (Poincaré) length of each side of B_i and B_{ij}, and we can show that the diameters of B_i and B_{ij} are bounded by $\pi/2 + \log 3$. Also we can show that the areas of B_i and B_{ij} are bounded by $5\pi/12$.

Step 3. *Construction of an auxiliary function.* Let

$$L_i = [L^{\alpha_i}] \quad \text{for } i = 1, 2, \cdots, h,$$

and

$$\gamma = \mu + \max_i \{\alpha_i + \rho_i'\} + \beta - \sum_i \alpha_i. \tag{3.23}$$

Lemma 3.4. *Let* L *be a sufficiently large integer of the form* (3.17), *and let* S *be the subset of* Γ_L *defined by Lemma* 3.3. *Then there exists a nonzero polynomial in* $\mathbf{Z}[X_1, \cdots, X_h]$,

$$P(X_1, \cdots, X_h) = \sum_{\ell_1=0}^{L_1} \cdots \sum_{\ell_h=0}^{L_h} a_{\ell_1 \cdots \ell_h} X_1^{\ell_1} \cdots X_h^{\ell_h}, \tag{3.24}$$

such that

$$\max \{|a_{\ell_1 \cdots \ell_h}|\} < e^{c_4 L^\gamma}, \tag{3.25}$$

and such that the function

$$F(z) = P(f_1(z), \cdots, f_h(z))$$

satisfies

$$F(w) = 0 \quad \textit{for all} \quad w \in S. \tag{3.26}$$

Proof. Since card $S = [L^\beta]$, the condition (3.26) is given by a system of $[L^\beta]$ linear homogeneous equations in $\prod_{i=1}^{h} ([L^{\alpha_i}] + 1)$

unknowns $a_{\ell_1\cdots\ell_h}$. For each w, the coefficients of the corresponding equation are in K_w. By (iii) of Theorem 1,

$$f_1(w)^{\ell_1}\cdots f_h(w)^{\ell_h} \prod_{i=1}^{h} \delta_{iw}^{L_i} \in I_{K_w},$$

and by (ii) and (iii) of Theorem 1, the maximum of the absolute values of its conjugates is at most $e^{2C'hL^{\max\{\alpha_i+\rho_i'\}}}$. Further we have $[K_w:\mathbf{Q}] \leq C'L^{\mu}$ by (i) of Theorem 1, and $\sum_i \alpha_i > \mu + \beta$ by (3.10). To solve the system of these equations, we use Siegel's lemma as usual. See for example Exercise 1.3.b of [2]. It is a generalization of Lemma 1.3.1 of [2] to the case where the number field containing the coefficients varies for each equation. Briefly speaking, we use the formula $\max_{1\leq i\leq N} |x_i| \leq (\sqrt{2}NA)^{\frac{\delta M}{N-\delta M}}$, where A is the maximum of the houses (i.e. $|\overline{}|$) of coefficients, and δ is the maximum of the degrees of the fields. Then, we find a non trivial integer solution $\{a_{\ell_1\cdots\ell_h}\}$ such that

$$\begin{aligned}
&\log \max \{|a_{\ell_1\cdots\ell_h}|\} \\
&\leq \frac{C'L^{\mu}[L^{\beta}]}{\Pi([L^{\alpha_i}]+1) - C'L^{\mu}[L^{\beta}]} \times \\
&\quad \times \{\log\sqrt{2} + \sum \log([L^{\alpha_i}]+1) + 2C'hL^{\max\{\alpha_i+\rho_i'\}}\} \\
&< c_4 L^{\mu + \max\{\alpha_i+\rho_i'\} + \beta - \sum \alpha_i}.
\end{aligned}$$

So by (3.23) the proof is complete.

Remark 3.2. By (3.10) and (3.23) we have

$$\gamma < \max_i \{\alpha_i + \rho_i'\}. \tag{3.27}$$

Step 4. *Lower bound of* $F(w)$. We use an arithmetical argument. The following lemma 3.5 is well-known.

Lemma 3.5. *Let* K *be an algebraic number field with* $[K:\mathbf{Q}] = \delta$. *Then for any* $\xi \in K$,

$$\log |\xi| \geq -(\delta - 1) \log |\overline{\xi}| - \delta \log \text{den}\, \xi.$$

Lemma 3.6. *Let* $N \geq L$ *and* $w \in \Gamma_N$, *and suppose* $F(w) \neq 0$. *Then*

$$\log |F(w)| \geq -c_5 N^{\mu} \sum_i N^{\rho_i'} L^{\alpha_i}.$$

Proof. Clearly $F(w) \in K_w$, and by (3.24), (3.25) and (ii) of Theorem 1 we have

$$|\overline{F(w)}| \leq \Pi ([L^{\alpha_i}] + 1) \times e^{c_4 L^{\gamma}} \times \Pi\, e^{C' N^{\rho_i'}[L^{\alpha_i}]}.$$

By (iii) of Theorem 1 we have

$$\text{den}\, F(w) \leq \Pi\, e^{C' N^{\rho_i'}[L^{\alpha_i}]}.$$

Also $[K_w:\mathbf{Q}] \leq C' N^{\mu}$. Then, noting $L >> 1$, we obtain the result by Lemma 3.5 and (3.27).

Step 5. *Upper bound of* $F(w)$.

Notation. For $N \geq L$, $R = 1 - N^{-(1+\varepsilon)}$, and $w \in D_N$, we put

$$\Lambda_N(w) = \sum_{w_\nu} \log |\phi_{R,w_\nu}(w)|,$$

where w_ν ranges over all zeros of F in $D_N - \{w\}$ counted with multiplicity, and

$$\phi_{R,w_\nu}(w) = \frac{R(w - w_\nu)}{R^2 - \overline{w}_\nu w}.$$

Let $\Phi = F \prod_{i=1}^{h} g_i^{L_i}$. Then Φ is holomorphic on D.

Lemma 3.7. *Let* $N \geq L$, $R = 1 - N^{-(1+\varepsilon)}$, *and* $w \in D_N$, *and suppose* $\Phi(w) \neq 0$. *Then*

$$\log |\Phi(w)| < c_4 L^{\gamma} + c_6 \sum_i N^{(1+\varepsilon)\rho_i} L^{\alpha_i} + \Lambda_N(w).$$

Proof. For $R = 1 - N^{-(1+\varepsilon)}$, $\log |g_i|_R$ and $\log |f_i g_i|_R$ are at most $C'N^{(1+\varepsilon)\rho_i}$ by (2) of Theorem 1. Hence by (3.24), (3.25), we obtain

$$\log |\Phi|_R < \sum \log ([L^{\alpha_i}] + 1) + c_4 L^\gamma + \sum C'N^{(1+\varepsilon)\rho_i} [L^{\alpha_i}]$$

$$\leq c_4 L^\gamma + c_6 \sum N^{(1+\varepsilon)\rho_i} L^{\alpha_i}.$$

Therefore by Jensen's formula we obtain

$$\log |\Phi(w)| < c_4 L^\gamma + c_6 \sum N^{(1+\varepsilon)\rho_i} L^{\alpha_i} + \sum_{w_\nu} \log |\phi_{R,w_\nu}(w)|,$$

where w_ν ranges over all zeros of Φ in $\{|z| \leq R\} - \{w\}$. Now we observe that if w_ν is a zero of F, then it is a zero of Φ at least of the same order. We observe also that every term $\log |\phi_{R,w_\nu}(w)|$ is negative since $|\phi_{R,w_\nu}(w)| < 1$. Then we see that the inequality holds even if the sum over w_ν is replaced by $\Lambda_N(w)$. Hence the results follows.

Lemma 3.8. *Let* $N \geq L$, $R = 1 - N^{-(1+\varepsilon)}$, *and* $w \in \Gamma_N$, *and suppose* $F(w) \neq 0$. *Then*

$$\log |F(w)| < c_4 L^\gamma + c_7 \sum_i N^{(1+\varepsilon)\rho_i} L^{\alpha_i} + \Lambda_N(w).$$

Proof. Since $|g_i(w)| \geq e^{-C'N^{\rho_i}}$ by (2) of Theorem 1, we have $\Phi(w) \neq 0$, and

$$\log |\Phi(w)| \geq - \sum C'N^{\rho_i} [L^{\alpha_i}] + \log |F(w)|.$$

Then by Lemma 3.7, we obtain the result.

Combining Lemmas 3.6 and 3.8, and using (3.27), we obtain the following lemma.

Lemma 3.9. *Let* $N \geq L$, $R = 1 - N^{-(1+\varepsilon)}$, *and* $w \in \Gamma_N$, *and suppose* $F(w) \neq 0$. *Then*

$$- c_8 N^{\mu} \sum_i N^{\rho_i'} L^{\alpha_i} - c_7 \sum_i N^{(1+\varepsilon)\rho_i} L^{\alpha_i} < \Lambda_N(w). \qquad (3.28)$$

In particular, if $N = L$, *then*

$$- c_8 L^{\mu} \sum_i L^{\alpha_i+\rho_i'} - c_7 \sum_i L^{\alpha_i+(1+\varepsilon)\rho_i} < \Lambda_L(w). \qquad (3.29)$$

Now we define the following properties for $0 \leq j \leq m$, and $N \geq L$. Note that S_j are the subsets of Γ_L defined by Lemma 3.3.

$\mathscr{S}_j$: $F(w) = 0$ for all $w \in S_j$.

$\mathscr{T}_N$: $F(w) = 0$ for all $w \in \Gamma_N$.

Note that $F(w) = 0$ for all $w \in S$ by Lemma 3.4.

Step 6. *Vanishing of* F *on* S *implies* $\mathscr{S}_0$. We shall show that vanishing of F on S implies $\mathscr{S}_0$. To this aim, suppose there exists a point $w \in S_0$ such that $F(w) \neq 0$. Note that $S \subset S_0$. But, since $F(w_\nu) = 0$ for all $w_\nu \in S$, we have $w \notin S$. Now from (3.20), we have $d(w, w_\nu) < c_2$. Hence by (2.1) we have, for $R = 1 - L^{-(1+\varepsilon)}$,

$$\log |\phi_{R,w_\nu}(w)| < - e^{-(2+\varepsilon)c_2} = - c_9 < 0.$$

Note that every term of $\Lambda_L(w)$ is negative, and card $S = [L^\beta]$ by (3.22). Then, taking the sum in $\Lambda_L(w)$ over all $w_\nu \in S$, we obtain

$$\Lambda_L(w) < - c_9 [L^\beta].$$

On the other hand, we have

$$\beta > \mu + \alpha_i + \rho_i', \quad \alpha_i + (1 + \varepsilon)\rho_i$$

for $i = 1, 2, \cdots, h$ by (3.11) and (3.12). But since $L >> 1$, this implies a contradiction by (3.29). Therefore $F(w) = 0$ for all $w \in S_0$, so $\mathscr{S}_0$ holds, as desired.

Step 7. *The condition* $\mathscr{S}_j$ *implies* $\mathscr{S}_{j+1}$ *for* $0 \leq j \leq m - 1$. Let j be an integer with $0 \leq j \leq m - 1$, and suppose $\mathscr{S}_j$ holds. Suppose further $\mathscr{S}_{j+1}$ does not hold, i.e., there exists a point

$w \in S_{j+1}$ such that $F(w) \neq 0$. Since $S_j \subset S_{j+1}$, we have for all points w_ν of S_j, $d(w,w_\nu) < (1 - 1/2^{j+1}) \log L + c_1$ by (3.18). Hence by (2.1) we have, for $R = 1 - L^{-(1+\varepsilon)}$,

$$\log |\phi_{R,w_\nu}(w)| < - e^{-(2+\varepsilon)\{(1-1/2^{j+1})\log L + c_1\}}$$

$$= - c_{10}' \; L^{-(2+\varepsilon)(1-1/2^{j+1})}.$$

Note that $\operatorname{card} S_j > c_3' L^{\kappa-1/2^j}$ for $0 \leqq j \leqq m - 1$ by (3.19) and (3.21). Then, taking the sum in $\Lambda_L(w)$ over all $w_\nu \in S_j$, we obtain

$$\Lambda_L(w) < - c_3' L^{\kappa-1/2^j} \; c_{10}' L^{-(2+\varepsilon)(1-1/2^{j+1})}$$

$$= - c_{10} L^{\kappa-1/2^j-2+1/2^j-\varepsilon(1-1/2^{j+1})}$$

$$< - c_{10} L^{\kappa-2-\varepsilon}.$$

On the other hand, we have

$$\kappa - 2 - \varepsilon > \mu + \alpha_i + \rho_i', \quad \alpha_i + (1 + \varepsilon)\rho_i$$

by (3.13) and (3.14). But this implies a contradiction by (3.29). Therefore $F(w) = 0$ for all $w \in S_{j+1}$, so $\mathcal{S}_j$ implies $\mathcal{S}_{j+1}$, as desired.

Step 8. *The condition $\mathcal{S}_m$ implies $\mathcal{J}_L$.* Suppose $\mathcal{S}_m$ holds. Suppose further $\mathcal{J}_L$ does not hold, i.e., there exists a point $w \in \Gamma_L$ such that $F(w) \neq 0$. Since the diameter of D_L is equal to $\log L + \log (2 - 1/L)$, we have, for $w_\nu \in S_m$,

$$d(w,w_\nu) < \log L + \log 2.$$

Hence by (2.1) we have, for $R = 1 - L^{-(1+\varepsilon)}$,

$$\log |\phi_{R,w_\nu}(w)| < - e^{-(2+\varepsilon)(\log L + \log 2)} = - c_{11}' L^{-2-\varepsilon}.$$

Note that card $S_m > CL^{\kappa-1/2^m} - 2$ by (3.19). Then, taking the sum in $\Lambda_L(w)$ over all $w_\nu \in S_m$, we obtain

$$\Lambda_L(w) < -(CL^{\kappa-1/2^m} - 2)\, c'_{11} L^{-2-\varepsilon} < -c_{11} L^{\kappa-2-2\varepsilon},$$

by (3.4). On the other hand, we have

$$\kappa - 2 - 2\varepsilon > \mu + \alpha_i + \rho'_i, \quad \alpha_i + (1+\varepsilon)\rho_i$$

by (3.13) and (3.14). But this implies a contradiction by (3.29). Therefore $\mathcal{S}_m$ implies $\mathcal{J}_L$, as desired.

Step 9. *The condition* $\mathcal{J}_N$ *implies* $\mathcal{J}_{N+1}$ *for* $N \geqq L$. Let N be an integer with $N \geqq L$, and suppose $\mathcal{J}_N$ holds. Suppose further $\mathcal{J}_{N+1}$ does not hold, i.e., there exists a point $w \in \Gamma_{N+1}$ such that $F(w) \neq 0$. Since the diameter of D_{N+1} is equal to $\log(N+1) + \log(2 - (N+1)^{-1})$, we have, for $w_\nu \in \Gamma_N$,

$$d(w, w_\nu) < \log(N+1) + \log 2.$$

Hence, similarly to Step 8, we have for $R = 1 - (N+1)^{-(1+\varepsilon)}$,

$$\log |\phi_{R,w_\nu}(w)| < -c'_{11}(N+1)^{-2-\varepsilon}.$$

Note that card $\Gamma_N \geqq CN^\kappa$ by (1) of Theorem 1. Then, taking the sum in $\Lambda_{N+1}(w)$ over all $w_\nu \in \Gamma_N$, we obtain

$$\Lambda_{N+1}(w) < -CN^\kappa c'_{11}(N+1)^{-2-\varepsilon} < -c_{12}(N+1)^{\kappa-2-\varepsilon}.$$

On the other hand, we have

$$\kappa - 2 - \varepsilon > \mu + \alpha_i + \rho'_i, \quad \alpha_i + (1+\varepsilon)\rho_i$$

by (3.13) and (3.14). But this implies a contradiction by (3.28). Therefore $\mathcal{J}_N$ implies $\mathcal{J}_{N+1}$, as desired.

Consequently, we obtain by induction the following result.

Lemma 3.10. *The property* $\mathcal{J}_N$ *holds for all* $N \geqq L$.

Step 10. *The function* F *vanishes identically*. Lemma 3.10 implies $F \equiv 0$ as follows. Let $w \in D$, and suppose $\Phi(w) \neq 0$. Also let $N >> L$. Since the Poincaré radius of D_N is smaller than $\frac{1}{2}(\log N + \log 2)$, we have, for $w_\nu \in D_N$,

$$d(w,w_\nu) \leqq d(0,w_\nu) + d(0,w) < \frac{1}{2} \log N + c'(w),$$

with $c'(w) = \frac{1}{2} \log 2 + d(0,w)$. Therefore by (2.1) we obtain, for $R = 1 - N^{-(1+\varepsilon)}$,

$$\log |\phi_{R,w_\nu}(w)| < - e^{-(2+\varepsilon)\{(\log N)/2 + c'(w)\}} < - c(w)N^{-1-\varepsilon/2}$$

with a constant $c(w)$ depending only on w. Since $\mathcal{J}_N$ holds by Lemma 3.10, we have $F(w_\nu) = 0$ for all $w_\nu \in \Gamma_N$. Also we have $\operatorname{card} \Gamma_N \geqq CN^\kappa$ by (1) of Theorem 1. Hence we obtain by Lemma 3.7

$$\log |\Phi(w)| < c_4 L^\gamma + c_6 \sum_i N^{(1+\varepsilon)\rho_i} L^{\alpha_i} - CN^\kappa c(w) N^{-1-\varepsilon/2}.$$

Then this inequality yields a contradiction by letting N to infinity, since $\kappa - 1 - \varepsilon/2 > (1 + \varepsilon)\rho_i$ by (3.14). Therefore $\Phi(w) = 0$ for all $w \in D$, and F is identically equal to zero. This contradicts the assumption of algebraic independence of $f_1,\cdots,f_h$. So the assumption (3.1) is false, and this concludes the proof of Theorem 2.

4. PROOF OF THEOREM 3

We assume that all the conditions of Theorem 3 are satisfied. We assume further that the conclusion of Theorem 3 is false, i.e.,

$$\kappa > 1 + \mu + \max \{\rho',\rho''\}, \tag{4.1}$$

and we shall show that this implies a contradiction. We may suppose μ, ρ' and ρ'' are positive.

We divide the proof into several steps.

Step 1. *Preliminaries*. The following auxiliary Lemmas (4.1, 4.2, and 4.3) are almost the same as Lemmas 5, 6, and 7 of [3], and are obtained by Leibniz's formula. Use the inequality

$$\sum_{k_1+\cdots+k_\ell=k, k_i\geqq 0} 1 \leqq (k+1)^{\ell-1}$$

for the proof, and if necessary, use further the inequalities $k \log (k+1) - k \leqq \log k! \leqq (k+1) \log (k+1) - k$ for nonnegative integers k. To prove Lemma 4.2, repeat the same argument as Lemma 4.1.

Lemma 4.1. *Let* f_1, f_2 *be functions satisfying* (ii) *of Theorem* 3. *Then for any* $\ell \geqq 1$, $k \geqq 0$, $N >> 1$, $w \in \Gamma_N$ *and* $i = 1,2$, *we have*

$$\left|\overline{(f_i^\ell)^{(k)}(w)}\right| \leqq e^{C'N^{\rho'}(\ell+k)} (k+1)^{C'N^{\rho''}k} (k+1)^\ell .$$

Lemma 4.2. *Let* f_1, f_2 *be functions satisfying* (ii) *of Theorem* 3. *Then for any* $\ell_1, \ell_2 \geqq 0$, $k \geqq 0$, $N >> 1$ *and* $w \in \Gamma_N$, *we have*

$$\left|\overline{(f_1^{\ell_1} f_2^{\ell_2})^{(k)}(w)}\right| \leqq e^{C'N^{\rho'}(\ell_1+\ell_2+k)} (k+1)^{C'N^{\rho''}k} (k+1)^{\ell_1+\ell_2} .$$

Lemma 4.3. *Let* f_1, f_2 *be functions satisfying* (iii) *of Theorem* 3. *Then for any* $\ell_1, \ell_2 \geqq 0$, $k \geqq 0$, $N >> 1$ *and* $w \in \Gamma_N$, *we have*

$$(f_1^{\ell_1} f_2^{\ell_2})^{(k)}(w) \prod_{i=1}^{2} \delta_{iw}^{\ell_i+k} (k!)^{\delta'_{iw}} \in I_{K_w},$$

and

$$\prod_{i=1}^{2} \delta_{iw}^{\ell_i+k} (k!)^{\delta'_{iw}} \leqq e^{C'N^{\rho'}(\ell_1+\ell_2+2k)} (k!)^{2C'N^{\rho''}} .$$

We introduce some quantities. Let ε be any positive number, and take any positive number α such that

$$\alpha > \mu + (1+\varepsilon)\rho + \varepsilon, \tag{4.2}$$

$$\alpha > 2\mu + \rho' + \varepsilon, \tag{4.3}$$

$$\alpha > \frac{\kappa - 1 + \mu}{2} + \varepsilon. \tag{4.4}$$

Set

$$\beta = 2\alpha - (\kappa - 1) - \mu - \varepsilon. \tag{4.5}$$

For simplicity, taking ε suitably we may suppose that α and β are rational. Further let us take any positive number β' such that

$$\beta' > \beta,$$
$$\kappa-3+\beta'-\varepsilon > \mu+\alpha+\rho',\ \mu+\beta+\rho',\ \mu+\beta+\rho'',\ \alpha+(1+\varepsilon)\rho. \tag{4.6}$$

These quantities satisfy the following relations which will be used later.

Lemma 4.4. *We have*

$$\beta > 0, \tag{4.7}$$
$$2\alpha > \kappa - 1 + \mu + \beta, \tag{4.8}$$
$$\kappa - 1 + \beta > \alpha + (1 + \varepsilon)\rho, \tag{4.9}$$
$$\kappa - 1 + \beta > \mu + \alpha + \rho', \tag{4.10}$$
$$\kappa - 1 + \beta - \frac{\varepsilon}{2} > (1 + \varepsilon)\rho. \tag{4.11}$$

Proof. Clearly (4.7) follows from (4.4). (4.8) follows from (4.5). By (4.5), (4.2) we have

$$\kappa - 1 + \beta - \alpha - (1 + \varepsilon)\rho = \alpha - \mu - (1 + \varepsilon)\rho - \varepsilon > 0,$$

so (4.9) holds. By (4.5), (4.3), we have

$$\kappa - 1 + \beta - \mu - \alpha - \rho' = \alpha - 2\mu - \rho' - \varepsilon > 0,$$

so (4.10) holds. By (4.9), (4.2), we have

$$\kappa - 1 + \beta - \frac{\varepsilon}{2} > \alpha + (1 + \varepsilon)\rho - \frac{\varepsilon}{2} > (1 + \varepsilon)\rho,$$

so (4.11) holds. Thus the proof is complete.

In the following, we shall denote by $c_1, c_2, \cdots$, or $c_1', c_2', \cdots$, positive constants depending only on the quantities so far introduced (and not depending on L below). Let L be a sufficiently large integer. For simplicity we assume that L^α and L^β are integers, which is possible since we took rational α and β.

For the case of Gel'fond's method also, for an auxiliary function we shall take the usual one. But for a set at which we require the auxiliary function to vanish with high multiplicity, we shall take a special subset S_L of Γ_L as is given by Lemma 4.5 below.

Step 2. *Choice of subsets* S_N *of* Γ_N.

Lemma 4.5. *For every* $N \geq L$ *there exists a subset* S_N *of* Γ_N *such that*

$$\text{(Poincaré) diameter of } S_N < c_1, \tag{4.12}$$

and

$$\text{card } S_N = [c_2 N^{\kappa-1}]. \tag{4.13}$$

Proof. The (Poincaré) area of D_N is

$$\int_0^{2\pi}\int_0^{1-1/N} \frac{r\,dr\,d\theta}{(1-r^2)^2} = \left(\frac{N}{2-1/N} - 1\right)\pi.$$

So we can cover D_N by at most $c_1'N$ subregions, each having bounded area and bounded diameter, say smaller than c_1. Then by the assumption (1) of Theorem 3, one of these subregions, say A_N, contains at least $[CN^{\kappa}/c_1'N]$ points of Γ_N. So, for $c_2 = C/c_1'$, it is possible to choose $[c_2N^{\kappa-1}]$ points of Γ_N contained in A_N. Denote by S_N the set of these points. Then S_N satisfies (4.12) and (4.13), as desired.

Remark 4.1. An example of covering of D_N having the above properties can be constructed as follows. Let n be the integer with $2^{n-1} < N \leq 2^n$. Then D_N is covered by B_{ij} of Remark 3.1 with $i = 1,2,\cdots,n$ and $j = 1,2,\cdots,2^i$. The number of these (closed) subregions is $2 + 2^2 + \cdots + 2^n = 2(2^n - 1) < 4N$. Similary to Remark 3.1, the diameter of B_{ij} is bounded by $\frac{\pi}{2} + \log 3$, and the area of B_{ij} is bounded by $\frac{5\pi}{12}$.

Step 3. *Construction of an auxiliary function.* Let

$$L_1 = L_2 = L^{\alpha},$$

and

$$\gamma = \max\{\alpha+\rho', \beta+\rho', \beta+\rho''\} + \kappa - 1 + \mu - 2\alpha + \beta. \tag{4.14}$$

Lemma 4.6. *Let* S_L *be the subset of* Γ_L *defined by Lemma* 4.5. *Then there exists a nonzero polynomial in* $\mathbb{Z}[X,Y]$,

$$P(X,Y) = \sum_{\ell_1=0}^{L_1} \sum_{\ell_2=0}^{L_2} a_{\ell_1\ell_2} X^{\ell_1} Y^{\ell_2} \tag{4.15}$$

such that

$$\max \{|a_{\ell_1\ell_2}|\} < L^{c_3 L^{\gamma}}, \tag{4.16}$$

and such that the function

$$F(z) = P(f_1(z), f_2(z))$$

satisfies

$$F^{(k)}(w) = 0 \tag{4.17}$$

for all $w \in S_L$ *and* $k = 0,1,\cdots,L^{\beta}$.

Proof. Since card $S_L = [c_2 L^{\kappa-1}]$, the condition (4.17) is given by a system of $[c_2 L^{\kappa-1}](L^{\beta} + 1)$ linear homogeneous equations in $(L_1 + 1)(L_2 + 1)$ unknowns $a_{\ell_1\ell_2}$. For each w, the coefficients are in K_w. By Lemma 4.3 we have

$$(f_1^{\ell_1} f_2^{\ell_2})^{(k)}(w) \prod_{i=1}^{2} \delta_{iw}^{L_i+k} (k!)^{\delta'_{iw}} \in I_{K_w},$$

and by Lemmas 4.2 and 4.3, for $k \leq L^{\beta}$ the maximum of the absolute values of its conjugates is at most

$$e^{C'L^{\rho'}(2L^{\alpha}+L^{\beta})} (L^{\beta}+1)^{C'L^{\rho''}L^{\beta}} (L^{\beta}+1)^{2L^{\alpha}} \times e^{2C'L^{\rho'}(L^{\alpha}+L^{\beta})} ((L^{\beta})!)^{2C'L^{\rho''}}$$
$$< L^{c_3' L^{\max\{\alpha+\rho', \beta+\rho', \beta+\rho''\}}} .$$

Further we have $[K_w : \mathbb{Q}] \leq C'L^{\mu}$ by (i) of Theorem 3, and $2\alpha > \kappa - 1 + \mu + \beta$ by (4.8). Hence by Siegel's lemma, there exists a non trivial integer solution $\{a_{\ell_1\ell_2}\}$ such that

$$\log \max \{|a_{\ell_1\ell_2}|\}$$

$$\leq \frac{C'L^{\mu}[c_2L^{\kappa-1}](L^{\beta}+1)}{(L^{\alpha}+1)^2 - C'L^{\mu}[c_2L^{\kappa-1}](L^{\beta}+1)}$$

$$\times \{\log \sqrt{2} + 2\log(L^{\alpha}+1) + c_3'L^{\max\{\alpha+\rho',\beta+\rho',\beta+\rho''\}} \log L\}$$

$$< c_3L^{\max\{\alpha+\rho',\beta+\rho',\beta+\rho''\}+\kappa-1+\mu-2\alpha+\beta} \log L.$$

So by (4.14) the proof is complete.

Remark 4.2. By (4.5), (4.14), we have $\gamma < \max\{\alpha+\rho', \beta+\rho', \beta+\rho''\}$ Further by (4.10) we have $\alpha + \rho' < \kappa - 1 - \mu + \beta$, and by (4.1) we have $\beta + \max\{\rho',\rho''\} < \kappa - 1 - \mu + \beta$. So we have

$$\gamma < \max\{\alpha+\rho', \beta+\rho', \beta+\rho''\} < \kappa - 1 - \mu + \beta. \tag{4.18}$$

Step 4. *Lower bound of* $F^{(k)}(w)$.

Lemma 4.7. *Let* $N \geqq L$ *and* $w \in \Gamma_N$, *and let* $\mathrm{ord}_w F = k \geqq 0$, *where* $\mathrm{ord}_w F$ *is the order of zero of* F *at* w. *Then*

$$\log |F^{(k)}(w)| \geqq - c_4N^{\mu}\{N^{\rho'}(L^{\alpha} + k) + N^{\rho''}k \log (k + 1)$$
$$+ L^{\alpha} \log (k + 1) + L^{\gamma} \log L\}.$$

Proof. We use an arithmetical argument. Clearly $F^{(k)}(w) \in K_w$. By (4.15), (4.16) and Lemma 4.2, we have

$$\overline{|F^{(k)}(w)|} \leq (L^{\alpha} + 1)^2 \times L^{c_3L^{\gamma}}$$
$$\times e^{C'N^{\rho'}(2L^{\alpha}+k)} (k + 1)^{C'N^{\rho''}k} (k + 1)^{2L^{\alpha}}.$$

By Lemma 4.3, we have

$$\mathrm{den}\,(F^{(k)}(w)) \leqq e^{2C'N^{\rho'}(L^{\alpha}+k)} (k!)^{2C'N^{\rho''}}.$$

Also $[K_w:\mathbb{Q}] \leqq C'N^{\mu}$. Then, noting $L >> 1$, we obtain the result by Lemma 3.5.

<u>Step 5</u>. *Upper bound of* $F^{(k)}(w)$. We use the same notation $\Lambda_N(w)$ as in Section 3, Step 5. We put $\Phi = F \prod_{i=1}^{2} g_i^{L_i}$. Then it is holomorphic on D.

<u>Lemma 4.8</u>. *Let* $N \geqq L$, $R = 1 - N^{-(1+\varepsilon)}$, *and* $w \in D_N$, *and let* $\operatorname{ord}_w \Phi = k \geqq 0$. *Then*

$$\log |\Phi^{(k)}(w)| < c_3 L^{\gamma} \log L + c_5 N^{(1+\varepsilon)\rho} L^{\alpha} + \Lambda_N(w) + k(\log N + c_6) + k \log (k+1).$$

<u>Proof</u>. For $R = 1 - N^{-(1+\varepsilon)}$, $\log |g_i|_R$ and $\log |f_i g_i|_R$ are at most $C'N^{(1+\varepsilon)\rho}$ by (2) of Theorem 3. Hence by (4.15), (4.16), we obtain

$$\log |\Phi|_R < 2 \log (L^{\alpha} + 1) + c_3 L^{\gamma} \log L + 2C'N^{(1+\varepsilon)\rho} L^{\alpha}$$
$$\leqq c_3 L^{\gamma} \log L + c_5 N^{(1+\varepsilon)\rho} L^{\alpha}.$$

Therefore by Jensen's formula we obtain

$$\log |\Phi^{(k)}(w)| < c_3 L^{\gamma} \log L + c_5 N^{(1+\varepsilon)\rho} L^{\alpha} + \sum_{w_\nu} \log |\phi_{R,w_\nu}(w)| - k \log \frac{R^2 - |w|^2}{R} + \log k!,$$

where w_ν ranges over all zeros of Φ in $\{|z| \leqq R\} - \{w\}$. But if w_ν is a zero of F, then it is a zero of Φ of at least the same order. So we can replace the sum over w_ν by $\Lambda_N(w)$. Further we have $\log (R/(R^2 - |w|^2)) \leqq \log N + c_6$. Also $\log k! \leqq k \log (k+1)$. Hence the result follows.

<u>Lemma 4.9</u>. *Let* $N \geqq L$, $R = 1 - N^{-(1+\varepsilon)}$, *and* $w \in \Gamma_N$, *and let* $\operatorname{ord}_w F = k \geqq 0$. *Then*

$$\log |F^{(k)}(w)| < c_3 L^{\gamma} \log L + c_7 N^{(1+\varepsilon)\rho} L^{\alpha} + \Lambda_N(w) + k(\log N + c_6) + k \log (k+1).$$

Proof. Since $|g_i(w)| \geqq e^{-C'N^\rho}$ by (2) of Theorem 3, we have

$$\Phi^{(k)}(w) = F^{(k)}(w) \prod_i g_i(w)^{L_i} \neq 0,$$

and

$$\log |\Phi^{(k)}(w)| \geqq -2\, C'N^\rho\, L^\alpha + \log |F^{(k)}(w)|.$$

Then by Lemma 4.8, we obtain the result.

Combining Lemmas 4.7 and 4.9, we obtain the following Lemma.

Lemma 4.10. *Let* $N \geqq L$, $R = 1 - N^{-(1+\varepsilon)}$, *and* $w \in \Gamma_N$, *and let* $\mathrm{ord}_w F = k \geqq 0$. *Then*

$$\begin{aligned}&- c_8 N^\mu \{N^{\rho'}(L^\alpha + k) + N^{\rho''} k \log(k+1) + L^\alpha \log(k+1) + L^\gamma \log L\}\\ &\quad - c_7 N^{(1+\varepsilon)\rho}\, L^\alpha\\ &< \Lambda_N(w).\end{aligned}$$

In particular, if we put $k = N^t$ $(t \geqq 0)$, *then*

$$\begin{aligned}&- c_8 N^\mu \{N^{\rho'}(L^\alpha + N^t) + N^{\rho''} N^t \log(N^t+1) + L^\alpha \log(N^t+1) + L^\gamma \log L\}\\ &\quad - c_7 N^{(1+\varepsilon)\rho}\, L^\alpha\\ &< \Lambda_N(w). \qquad (4.19)\end{aligned}$$

We define the following properties for $N \geqq L$. Note that β' is the number defined by (4.6) and S_N is the subset of Γ_N defined by Lemma 4.5.

$\mathcal{S}_N$: $\mathrm{ord}_w F \geqq N^\beta$ for all $w \in S_N$.

$\mathcal{S}'_N$: $\mathrm{ord}_w F \geqq N^{\beta'}$ for all $w \in S_N$.

$\mathcal{T}_N$: $\mathrm{ord}_w F \geqq N^\beta$ for all $w \in \Gamma_N$.

Note that $\mathcal{S}_L$ holds by Lemma 4.6.

Step 6. *The condition* $\mathcal{S}_N$ *implies* $\mathcal{S}'_N$. Let N be an integer with $N \geqq L$, and suppose $\mathcal{S}_N$ holds. Suppose further $\mathcal{S}'_N$ does not hold, i.e., $\min\{\mathrm{ord}_w F \mid w \in S_N\} = k = N^t$ with $t < \beta'$. By the assumption, $\beta \leq t$. Let $w \in S_N$ be a point such that $\mathrm{ord}_w F = k$. For other points w_ν of S_N we have $\mathrm{ord}_{w_\nu} F \geq k$. By (4.12), $d(w, w_\nu) < c_1$. Hence by (2.1) we have, for $R = 1 - N^{-(1+\varepsilon)}$,

$$\log |\phi_{R,w_\nu}(w)| < - e^{-(2+\varepsilon)c_1} = - c_9 < 0.$$

So, taking the sum in $\Lambda_N(w)$ over all $w_\nu \in S_N$ with $w_\nu \neq w$, we obtain by (4.13),

$$\Lambda_N(w) < - c_9([c_2 N^{\kappa-1}] - 1)N^t.$$

Since $\beta \leq t$ and $t < \beta'$ (i.e., t is bounded from above), this implies a contradiction by (4.19) and the inequalities

$$\kappa-1+t > \mu+\alpha+\rho',\ \mu+t+\rho',\ \mu+t+\rho'',\ \mu+\gamma,\ \alpha+(1+\varepsilon)\rho$$

(for the last inequalities, see (4.10), (4.1), (4.1), (4.18), (4.9) respectively). Therefore $t \geq \beta'$. Consequently $\mathscr{S}_N$ implies $\mathscr{S}'_N$, as desired.

Step 7. *The condition* $\mathscr{S}'_N$ *implies* $\mathscr{T}_{N+1}$; $\mathscr{T}_{N+1}$ *implies* $\mathscr{S}_{N+1}$. Let N be an integer with $N \geq L$, and suppose $\mathscr{S}'_N$ holds. Suppose further $\mathscr{T}_{N+1}$ does not hold, i.e., there exists a point $w \in \Gamma_{N+1}$ such that $\mathrm{ord}_w F = k = (N+1)^t$ with $t < \beta$. Since the diameter of $D_{N+1} = \log(N+1) + \log(2 - (N+1)^{-1})$, we have, for $w_\nu \in S_N$,

$$d(w, w_\nu) < \log(N+1) + \log 2.$$

Hence by (2.1) we have, for $R = 1 - (N+1)^{-(1+\varepsilon)}$,

$$\begin{aligned}\log |\phi_{R,w_\nu}(w)| &< - e^{-(2+\varepsilon)(\log(N+1)+\log 2)} \\ &= - c_{10}(N+1)^{-(2+\varepsilon)}.\end{aligned}$$

Since $\mathrm{card}\, S_N = [c_2 N^{\kappa-1}]$, the condition $\mathscr{S}'_N$ implies

$$\Lambda_{N+1}(w) < - c_{10}(N+1)^{-(2+\varepsilon)}([c_2 N^{\kappa-1}] - 1)N^{\beta'}.$$

On the other hand, by (4.6) we have the inequalities

$$\kappa-3+\beta'-\varepsilon > \mu+\alpha+\rho',\ \mu+\beta+\rho',\ \mu+\beta+\rho'',\ \alpha+(1+\varepsilon)\rho,$$

hence by the first inequality of (4.18) we obtain

$$\kappa-3+\beta'-\varepsilon > \mu+\alpha+\rho',\ \mu+\beta+\rho',\ \mu+\beta+\rho'',\ \mu+\gamma,\ \alpha+(1+\varepsilon)\rho,$$

and so by $t < \beta$,

$$\kappa-3+\beta'-\varepsilon > \mu+\alpha+\rho',\ \mu+t+\rho',\ \mu+t+\rho'',\ \mu+\gamma,\ \alpha+(1+\varepsilon)\rho.$$

Then we get a contradiction by (4.19). Therefore $t \geqq \beta$. Consequently $\mathcal{S}'_N$ implies $\mathcal{J}_{N+1}$, as desired.

The condition $\mathcal{J}_{N+1}$ implies $\mathcal{S}_{N+1}$ immediately since $S_{N+1} \subset \Gamma_{N+1}$. Thus by induction we obtain the following Lemma.

Lemma 4.11. *The property* $\mathcal{J}_N$ *holds for all* $N \geqq L$.

Step 8. *The function* F *vanishes identically.* Let $w \in D$, and suppose $\Phi(w) \neq 0$. Put $R = 1 - N^{-(1+\varepsilon)}$ for $N >> L$. Then, identically to Section 3, Step 10, there exists a positive constant $c(w)$ such that for $w_\nu \in \Gamma_N$ we have

$$\log |\phi_{R,w_\nu}(w)| < - c(w)\, N^{-1-\varepsilon/2}.$$

Since $\mathcal{J}_N$ holds by Lemma 4.11, we have $\mathrm{ord}_{w_\nu} \Phi \geqq N^\beta$. Then, viewing that $\mathrm{card}\, \Gamma_N \geqq CN^\kappa$ by (1) of Theorem 3, we obtain by Lemma 4.8 with $k = 0$,

$$\log |\Phi(w)| < c_3 L^\gamma \log L + c_5 N^{(1+\varepsilon)\rho} L^\alpha - c(w) N^{-1-\varepsilon/2}\, CN^\kappa N^\beta.$$

Then this inequality yields a contradiction by letting N to infinity, since $\kappa - 1 + \beta - \varepsilon/2 > (1 + \varepsilon)\rho$ by (4.11). Therefore $\Phi(w) = 0$ for all $w \in D$, and F is identically equal to zero. This contradicts the assumption that f_1 and f_2 are algebraically independent. So the assumption (4.1) was false, and this concludes the proof of Theorem 3.

REFERENCES

[1] Gramain, F., Mignotte, M. and Waldschmidt, M.: Valeurs algébriques de fonctions analytiques, Acta Arith., 47, 97-121 (1986).

[2] Waldschmidt, M., Nombres Transcendants, Lecture Notes in Math., 402, Springer-Verlag, Berlin-Heidelberg-New York, 1974.

[3] Waldschmidt, M.: On functions of several variables having algebraic Taylor coefficients, Transcendence theory: Advances and applications, Academic Press, London, 169-186 (1977).